SOIL AND ORCHARD MANAGEMENT

SOIL AND ORCHARD MANAGEMENT

Dr. K.K. Sharma
Senior Horticulturist (Retd.)
Department of Horticulture,
P.A.U. Ludhiana

Ex. Director,
Fruit Research Station
Gangian (Dasuya)
Hoshiarpur (P.A.U.)

Dr. Nav Prem Singh
Asstt. Professor of Horticulture
Deptt. of Horticulture
P.A.U. Ludhiana

2026
Daya Publishing House®

A Division of
Astral International Pvt. Ltd.
New Delhi – 110 002

First Published, 2011

Reprinted, 2026

ISBN 978-81-7035-685-1 (PB)

Published by: **Daya Publishing House**
A Division of
Astral International Pvt. Ltd.
– ISO 9001:2015 Certified Company –
4736/23, Ansari Road, Darya Ganj
New Delhi-110 002
Ph. 011-43549197, 23278134
E-mail: info@astralint.com
Website: www.astralint.com

Laser Typesetting : **Twinkle Graphics**
Delhi - 110 088

Printed at : Replika Press Pvt. Ltd.

Foreword

In India many kinds of fruits are being cultivated for centuries. However, in North India plains tropical, sub-tropical and low chilling requiring cultivars of temperate fruits are being cultivated on a large scale. Over the years the emphasis on diversification is being given and with the availability of quality nursery plants the area under fruits has increased and is likely to increase further. Due to the availability of effective scientific production technology the fruit production per unit area has increased. Postharvest management of fruits have also helped the growers in increasing the profits from fruit-growing.

The consumption of fruits per person has also increased due to education. To meet the growing demand for fruits at affordable rates, the area under fruits have to be increased further.

Fruit growing is a long term proposition and require adequate knowledge of soil and plant physiology. Fruit production technology has become highly specialized. To produce high quality fruits for a number of years, from an orchard, with the use of inputs at economical rates is need of the time.

Dr. K.K. Sharma and Dr. Nav Prem Singh have done a good job to consolidate the desired information in this book, which will provide the orchardists efficient and cost-effective production technology in simple language.

This book shall be of immense value to the students, horticulturists, extension workers and orchardists, particularly of North Indian plains.

Dr. S.K. Salwan
Ex-Vice Chancellor
Punjab Technical University, Jalandhar.

Preface

Wide spread vitamin deficiencies and malnutrition in people can only be managed by increasing production of fruits per unit area. This increase can be brought about by the use of efficient and effective orchard management technology. The deterioration in the soil health due to the excessive use of synthetic fertilizers and environmental pollution caused by over use of pesticides should be discouraged. The judicious use of these ingredients can go a long way in increasing the orchard income.

The chapters dealing with soil and irrigation management shall help in improving soil health and check depletion of under ground water. The importance of organic matter in checking the micro-nutrient deficiencies in fruit trees can not be ignored. Chapters on weed management and other cultural practices shall help in the reduction of cost of fruit cultivation.

The recommendations made in this book for the management of pests and control of diseases in various fruits have been tested at the growers fields and shall prove very useful to the orchardists. The technique to rejuvenate old declining orchards shall help the growers in enhancing the productive life of their orchards. The information on frost and postharvest fruit handling shall help the orchardists to increase their profits.

I have relied on my own research work and field experience in the completion of this book. Hence, the field tested orchard management practices in each fruit shall be very useful to the extension worker and fruit growers in improving the quality of produce and long life of the orchard. The students, scientists and teachers shall be immensely benefited. The book shall form the basis of future research in the field of Horticulture.

I am grateful to Dr. Karaj Singh Dhillon, Senior Soil Scientist; Dr. B.S. Gill, Senior Agronomist; and Dr. B.V.C. Mahajan, Horticulturist for making valuable suggestions and rendering help in the improvement of respective chapters of their field. I am also indebted to Dr. Shivinder Singh Sandhu and his Orchard Manager Jasminder Singh Bhinder and Dr. H.S. Cheema for motivating us to write this book. I am grateful to my wife Sneh Lata Sharma and son Jagdeep Sharma for helping me at home front, without their help this job could not be completed.

Dr. K.K. Sharma

Dr. Nav Prem Singh

Contents

1

SOIL MANAGEMENT

Soil is the biologically modified weathered mantle. It is an essential component of ecosystems sustaining all, living vegetation, microorganisms, herbivores and carnivores. Soil is outermost layer of the earth. Soil productivity is the quality of the soil to produce crops with a set of management practices. Soil is one of the major factor which determines the success of an orchard. Since it is composed of all those materials required for growth and development of fruit crops. The materials present in the soil are in solid, liquid and gaseous states. These materials vary in their percentages from soil to soil. The mineral matter (clay, silt and sand), water, air and organic matter constituents determine the quality of a soil. While planting an orchard, whole soil profile is taken into consideration. Digging a pit shows the different layers in a soil profile. Usually, these are nearly the same and are not distinct in many areas. Air and water percentages vary significantly with soil texture and weather. Soil texture affects nearly every aspect of soil use and management. The soil types can be determined on the basis of the proportion of clay, silt and sand particles present in a given soil. The soil classification based on particle size is termed as textural classification of soil.

Particle	*Particle Size (mm)*
Clay	0.001-0.002
Silt	0.002-0.05
Sand	0.15-1.00

On the basis of percentages of these particles in a soil, the kind of soil can be easily determined. Accordingly soils are classified as clay loams, silt loams, sandy loams and sand. Sandy loam soil may consist of 75% sand, 14% silt and 11% clay particles. A clay loam shall have 34% sand, 39% silt and 27% clay. On the other hand soil structure refers to the arrangement of these particles in a soil.

The soil particles are held together in aggregates of various sizes and shapes. Maintenance of fine granular soil structure is of prime importance. Soil organic matter though present in small amount (less than 5%) but is utmost important component. Soil organic matter strongly modifies soil organism habitat. Its decomposition produces nitrogen, phosphorus and sulphur. Decomposition of organic matter is a continuous–process. On the basis of percentage and level of decomposition of organic matter in a particular soil can be classified as muck or peat. Muck soils are the most desirable ones for growing of horticultural crops. These soils are of three kinds depending upon the soil reaction.

1. Alkaline soils
2. Acidic soils
3. Moderately acidic soils.

The nature of the soil can be determined from the soil testing laboratory. For vegetables and cereal, only top 30cm of soil depth is taken for testing. Top soil has more organic matter than deeper layers of soil. The decay of organic residues by the activity of earthworms and other microorganisms has made the top soil fertile than the sub-soil. Orchardists are not concerned with the nature of the top soil alone. Hence, for fruit plantations soil is analyzed up to two meters depth, since, roots penetrate deep in the subsoil.

Soil Testing

Soil is a natural gift. Fruit tree gets nutrition through feeder root system from the soil. Root penetration in the soil and fertility level of the sub-soil determines the availability of nutrients to the fruit trees. There is a direct relationship between the concentration of nutrients in a particular soil and uptake by the fruit plants. It is not as simple as is expected from a given soil.

The presence of salts in soil solution and soil type affects the removal of nutrients by the fruit plant; from the soil, even if these are present in sufficient amounts. Moreover, the growth of fruit trees is greatly affected by total pore space, which is less in sandy soils and more in clays. Sandy soils are more porous than clays due to more macro-pores in sands. The soils in north India are not uniform due to the their alluvial nature. The nature of these soils varies from place to place. It becomes difficult to get representative soil sample for planting an orchard. Moreover, only 500 grams of composite soil sample is supplied to the soil testing laboratory. Seven such samples from top 0-15cm, 15-30cm, 30-60cm, 60-90cm, 90-120cm, 120-150cm and 150-200cm soil depth are separately tested. From each soil sample only few grams of soil is used for analysis. That very small quantity is considered as a representative of the whole area, which is to be brought under fruit tree plantation. This forms the major drawback in assessing the suitability of soil for a particular fruit plant. To minimize the error, samples from each acre should be taken separately. Label each bag indicating the depth of the layer from which sample has been taken. Send the samples to soil testing laboratory.

Suitability Limit of Soil for Orchard plantation

Character	*Citrus and Peach*	*Other fruit Plants*
I. Conductivity (mm hos/cm) or ds/m.	0.5	1.0
II. Calcium carbonate (%)	5	10
III. Lime concentration (%)	10	20
IV. pH	8.5	up to 8.7
	if EC is below 0.5 mm hos/cm.	

Alluvial Soils

These soils have developed by the deposition of silt by the flow of river waters during rainy season. These soils have low organic matter. The soils vary from neutral to alkaline in soil reaction. These soils develop a *canker* pan due to high lime content. Due to this hard pan the soil become saline-alkaline.

Increasing population pressure and decreasing cultivable land has compelled the growers to put all kinds of soil under cultivation. In North India alluvial as well as arid soils and soils in river basins are being brought under fruit plants. However, an ideal soil for growing fruit trees should be deep and well drained. The soil should have sufficient water and nutrient holding capacity. The pH of soil may vary between 6.5-8.5. Moreover, it should be free from hard pan and salt concentrations.

It is praiseworthy to the North Indian farmers, who are planting orchards judiciously in the river basins and foot hills. In the river basins there is no dearth of water but soil is not uniform hence, the plants of guava and pear can tilt in such a plantation, due to the large size of trees. On the other hand foot hill soils get depleted of nutrients and organic matter due to erosion for decades. The bunding and water conservation coupled with growing of legumes in newly planted orchards is helping the growers in profuse growth of citrus, mango, litchi and aonla plantations.

Importance of Soil Organic Matter

Organic matter influences the soils in various ways. The presence of sufficient amount of organic matter, effect the physical properties of the orchard soil, there is improvement in the soil granulation, water holding capacity and reduction in plasticity and cohesion. In the top layer of soil some partially decayed plant and animal residues accumulate. The earthworms and other microorganisms attack materials and mix it with the soil constituents. Amount of organic matter varies from area to area (0.5% to 5.0%). Earthworms are most important macro animals of the soil. They pass tons of dry earth through their bodies annually. Organic matter and mineral constituents of soil are subjected to grinding and digestive enzymes of the earth-worms. Thus these organisms are helpful in increasing the soil organic matter, nitrogen, calcium, magnesium, phosphorus and potash. Earthworms increase soil aeration and drainage. These are found mostly in medium to heavy soil, where moisture holding capacity is high along with high amount of organic matter. Their number is also higher in orchards where farm yard manure application and green manuring is usually carried out. The orchards where glyphosate has been used to control weeds, the activity of earthworms get reduced. The fruit trees like mango, litchi, pear, guava, ber and peaches etc. have huge root system. The rootlets grow fast in deeper layers of soil for search of water and nutrients. Many rootlets may die in course of time. They continually contribute for microbial activity. That is why; orchards and forestry plantation have higher activity of earthworms than fields where crops like wheat and paddy are grown. This may be due to yearly burning of trash and soil application of weedicides to control weeds in crops.

The newly planted orchards and those who are planning to adopt fruit cultivation, shall face a peculiar problem of micronutrient deficiencies, due to decreased organic matter content. The burning of trash every year has led to the loss of organic matter, decreased availability of essential nutrients and decreased rate of water percolation. It is necessary to add organic matter to soils. Green manuring and growing of legumes as intercrops in orchards during juvenile period of fruit plants shall be beneficial to the soil. It will maintain the optimum level of organic matter and shall subsequently improve soil health. The increased activity of microorganisms in soil shall be mutually beneficial. Some groups of microorganism produce products which are utilized by other groups as food. The activity of microorganisms influences the growth of fruit trees. The microbial activity in soil produces certain organic and inorganic acids which act as solvents of minerals. For example acids like nitric acid and diluted sulphuric acid affect the availability of phosphorus, iron and manganese. The presence of nitrate and ammonium compounds, which has been transformed from protein and related compounds by some changes in the soils are used by the fruit trees. Hence, to get good growth of fruit trees it is important to improve the level of organic matter in the orchard soils. Since, decomposition of organic matter is continuous process, therefore it should be added to the orchard soil continuously through various means.

Soil pH

The performance of fruit plant species greatly depend upon the reaction of the soil solution. The soils can be acidic, neutral or alkaline in nature. Choice of planting an orchard with the kind of fruit plant depends much upon the soil pH. Soil acidity is common where sufficient leaching of exchangeable bases from surface layer of soil have taken place. Alkaline conditions occur where bases have accumulated. The presence of calcium, magnesium and sodium carbonates increases OH^- and H^+ ions in soil solution. Alkaline soils are prevalent in central districts and irrigated arid zone of Punjab. Soils become acidic (low pH), if in a given soil micelle the adsorbed hydrogen is higher, which result into high H^+ ions concentration in the soil solution. The alkaline soil (high pH) is the result of hydrolysis of soil colloids saturated with calcium, magnesium and potassium cations. The concentration of H^+ ions in the soil solution decreases and simultaneously OH^- ions concentration will increase. The soil solution will turn alkaline in reaction. The pH of the soil solution depends upon there relative amounts of adsorbed hydrogen and adsorbed metallic cations. The pH of soil do not remain constant. Rains, irrigation water, addition of manures and fertilizers affect the soil pH. Most of the North Indian soils are slightly to moderately calcareous in nature. Only some soils in foot hills are noncalcareous and have neutral or slightly to moderately high soil pH. Most of the soils in arid irrigated areas of north Indian plains have slightly to moderately high soil pH (alkalinity). These soils can be judiciously used for the cultivation of fruit plants. The soils which are both calcareous and strongly alkaline can be put under ber, guava, pomegranate and aonla fruits. The farmer should know before hand the soil pH and other characters prior to the plantation of an orchard.

Now, the farmers are planning to plant orchards in the Shivalik foot hills. Here the land is plenty but undulated with small ravines. These soils are affected from moderate to high water erosion due to their topography. Due to high erosion of soil most of the soils are sandy to coarse loams in nature and have pH in the range of 7.1 to 8.5. These soils are most suited for orcharding particularly guava, citrus, chiku, mango, aonla and litchi plantations. Intensive soil conservation measures need to be adopted to check soil erosion and conserve rain water. Kacha dams have been built through out the kandi belt to store rain water for use as per requirement. Near the 'Choes' 'babbar', grass should be grown, which can be used for paper making and is helpful in checking the flow of rain water.

Good soil management by improving organic matter content and physical condition of soil shall determine the longevity and productivity of the fruit trees. The over use inorganic and synthetic fertilizers in orchards is unscientific and unethical. It affects the health of the orchard soil adversely and subsequently fruit trees get affected. To improve the soil organic matter content of orchard soil, in addition to ordinary farm yard manure, the vermicompost, green manuring and other cultural practices needs to be adopted. These measures are discussed below.

Green Manuring

Cultivation of green crops for two to three years prior to the planting of an orchard is advantageous for the fruit crop. These crops can also be grown after planting of an orchard. Green manuring crops should be grown with the application of additional fertilizers to reduce the competition between the fruit crop and green manuring crop. Green manuring with leguminous crops is more advantageous than ordinary crop. The succulent legumes *viz* dhaincha, sunhemp, guara, peas, cowpeas, soybean, senji, barseem, arhar or moong can be selected depending upon the climate and time of sowing prevailing in an area. Sometimes mixed green manures of legume and non-legume crop is grown to produce large amount of green material. Non legume crops can provide physical support to the legume crops which otherwise tend to lodge on the ground due to weak stem. Non legumes are hardy crops and can be grown easily in light soils too. Such a crop can fit well in the rotation of intercrops. The mixtures are grown in vacant spaces between the fruit plants during rainy season. The green manuring crops should be ploughed down at mid growth, when it has attained sufficient height, but still soft for decomposition. The high amount of water in a succulent leguminous crops help in the easy and fast decay, which will provide immediate benefits to the fruit trees. The legumes should be preferred over the non-legume. The presence of root nodules on the roots of the legumes provides extra nitrogen. In areas where rain is scanty one should be extra cautious in carrying out green manuring, since sufficient moisture is required for its decay. In arid zones green manuring should only be done when plenty of water is available from canals or tube wells.

Legumes

A group of herbaceous plant along with some grasses are very useful in enriching the orchard soils. Leguminous and non-leguminous, nitrogen fixing species help in the reclamation of soil. The legumes form nodules in the root system in association with the soil-borne bacterium (Rhizobium spp.). In the nodules a close biochemical and biophysical relation develops between the plant and the bacterium. The nodules fix atmospheric nitrogen which is utilized by the legume plant. Due to this character of fixing atmospheric nitrogen legumes can grow easily in nitrogen deficient soils. It has been found that some legumes are capable of adding 30-110kg of nitrogen per hectare per year. Nitrogen fixation in soils is dependent upon many factors. The most important is efficiency of rhizobia in the environment where the legume crop is being grown. Second is the quantity of green mass being added.

The following table shows the amount of green vegetative matter and nitrogen supplied by some legumes per hectare.

Legume	*Green matter Quintals per ha.*	*Nitrogen kg per ha.*
Dhaincha (*Sesbania aculeate*)	144	77
Sun hemp (*Crotalaria juncea*)	152	84
Barseem (*Trifolium alexandrinum*)	111	60
Senji (*Melilotus parviflora*)	206	134
Guara (*Cyamopsis tetragonolobus*)	144	62
Moong (*Vigna radiata*)	57	37

The legumes require little or no N fertilization. Applications of small amount of nitrogenous fertilizer help in the fast growth of the legumes. It also improves the succulency and quantity of biomass per hectare. The energy from the sun which is utilized by the legumes through photosynthesis and nutrients essential for growth are stored in legume plants and are recycled through decomposition by micro and macroorganisms in soils. The soil organic matter formed during this process serves as a continuous nutrient supply base. The physical properties of soil get stabilized through organic matter. To maintain productivity of the fruit trees, soluble nutrients are taken from the soil. These amounts must be replenished to maintain plant growth and fruit production. Therefore, to provide nutrients, biological decomposition of plant residue is the major source. The microorganisms, which improve the fertility status of the soil and contribute to plant growth are known as 'biofertilizers'. Similarly, other soil microorganisms those produce compounds such as vitamins and plant hormones can improve plant health and contribute to higher fruit yield. These microorganisms are called 'Phytostimuldor'. Large number of pathogenic microorganisms are present in soil and may cause considerable damage to plant roots. Some antagonistic microorganisms to these pathogens can prevent the infection of fruit plants. This prevention is carried out by the production of secondary metabolites *viz* antimicrobial metabolites, antibiotics and extra cellular enzymes. Some other microorganisms in soils produce compounds that stimulate the natural defence mechanism and improve the tree resistance to pathogens. Collectively these soil microorganisms have been termed as biopesticides. The beneficial microorganisms when are inoculated to the legume seeds particularly barseem result in the improvement of plant growth and total mass for green manuring.

CROPS SUITABLE TO GREEN MANURING

Dhaincha (*Sesbania aculeate*)

It is suitable for orchards on heavy soils (loam-clay). It is resistant to drought as well as stagnation of water., Dhaincha grows well in alkaline soils and corrects alkalinity if repeatedly grown for 4-5 years. The roots have plenty of nodules. It yields 12-15 tonnes of green manure per hectare. A seed rate of 35-40 kg/ha will be sufficient. Ploughing may be done after 40-45 days of sowing. Application of 25kg diammonium phosphate at the time of sowing greatly improves the growth of dhaincha.

Sunhemp (*Crotalaria juncea*)

It is a quick growing green manure crop. It can be ploughed after 40-45 days of sowing. It can not withstand heavy irrigation. A good stand of crop may provide 13-16 tonnes of green biomass per hectare. Seed required for sowing one hectare is 30-35 kg.

Barseem (*Trifolium alexandrinum*)

Barseem is grown in almost all kinds of fruit crops expect peaches. Two to three cuttings are taken before it is finally ploughed in the orchards. 18-20 kg of seed is required per hectare. It yields more than 10 tonnes of green biomass per hectare. In addition to green matter it adds sufficient beneficial microorganisms.

Senji (*Melilotus parviflora*)

It is an important green manure crop of orchards. It is usually sown in October-November and 20-25kg of seed is sufficient to sow one hectare. It can add more than 20 tonnes of green biomass. If seed ripens it sheds itself and grows of its own to a number years in orchards and add microbes through nodules.

Green manuring maintain the fertility of the orchard soil. The annual application of fertilizer doses to the fruit plants can be reduced to a great extent. Green manures check the leaching of nutrients by the addition of humus. Humus is highly colloidal and amorphous in nature. The cation exchange capacity and water holding capacity of soil with the increase of humus, increases to a great extent. Dependence on the farm yard manure shall be reduced, which is too short of the requirement. Ploughing a leguminous crop shall improve organic matter and nitrogen content of soil. It has been estimated that addition of two tonne of green manure dry matter per acre would add forty kilograms to fifty kilograms of nitrogen. Green manuring crop take up the soluble constituents from soil, those might otherwise have least through soil erosion or leaching. Green manuring crop also acts as a cover crop in orchards.

Soil Management in Orchards

Management of orchard floor does influence orchard productivity. Soil management under the fruit trees is very difficult, due to the various training systems followed in different fruit crops. It is little easy to work under the canopy of tree crops. There is no single system which can be followed for all soils and all fruit plants.

Objectives of Orchard Soil Management

1. To preserve soil structure
2. To control weeds
3. To improve soil organic matter
4. To improve irrigation facility to fruit plants
5. To facilitate movement in the orchard
6. To reduce the root diseases
7. To improve availability of nutrients
8. To save fruit plants from frost and freezes.

There are three options to achieve the above objectives. (*a*) Sod culture (*b*) Cover crops (*c*) Clean cultivation.

(*a*) Permanent or Partial Sod Culture

Permanent sod is a difficult proposition in orchards due to shading of the fruit trees and close plantation *e.g.* Phalsa or grapes. Due to shade permanent sod culture can not survive during winter months. The poor growth of sod is beneficial during summers, as sod do not compete for water and nutrients with the fruit plants. Permanent sod is also very difficult to preserve in orchards, which are flooding or segment irrigated. Sod checks the flow of water and huge quantity of water is wasted through percolation. Sod can be created in orchards which are under sprinkler or drip irrigated. These irrigation methods/systems do not irrigate entire vacant area, hence the sod species shall die as a result of drought.

Grasses and weeds do grow in orchards particularly in North Indian plains. These make lush growth during monsoons. Although weeds and grasses need to be removed from fruit tree basins to save water and nutrients, even then these do add some quantity of organic matter to the soil. Sod culture is not practiced in the younger planations due to growing of inter crops for early three to four years of planting an orchard. In later years if sod culture is applied in orchard soil it will be beneficial in checking loss of soil moisture and shall protect top soil from wind and water erosion. Plant roots holds the soil particles in place. In Europe, America and Australia sod culture is generally practiced in orchards. It will be useful in orchards planted in hilly areas on contours. In sod culture a thick carpet of grass is allowed to grow to cover the vacant area or a cover crop is grown. Occasionally the grass or cover crop is moved down in order to increase the organic matter content of the orchard soil and to decrease competition for moisture and nutrients. Very few orchards in India carry out sod culture, where as in Australia many citrus orchards are under permanent sod culture. In India too some kind of adjustment between clean cultivation and sod culture shall be advantageous. It has been proved at the farmers field that instead of having big fields of 20 to 40 acres of an orchard, one should have one acre segments with strong, wide ridges and big or small are advantageous. The Khabal grass should be allowed to grow on the paths, segments and vacant space out side the tree basins to from a sod. This allows rain water percolation and checks surface runoff. It also lowers the severe hot summer temperature of the orchard and lessens soil compaction.

In general, aim of the orchardist should be to provide conditions for the fruit trees to flourish on the available nutrition and moisture. The tree basins should be kept free of weeds.

Farm yard manure and fertilizers should be mixed with soil in the tree basins uniformly so that feeder roots can easily take up these nutrients from the soil solution. The cultural practices depend much upon the kind of fruit being grown.

(*b*) Cover Crops

Growing of cover crops in orchard is the only system of soil management in hands of the farmers. Many kinds of crops can be grown in orchards as cover crops. It is the cheapest and most effective method of maintaining soil productivity and soil structure. The orchards where some kind of cover crop (intercrop) have been grown, the soil is open and friable due to the addition of organic matter and root penetration. Thus the water holding and nutrient holding capacity is improved along with increased soil aeration. In arid irrigated zone of north India, it is beneficial because the soils are sandy and improvement of organic matter in such soils is advantageous to the orchards. Especially water remains in the upper root zone near the surface soil, which is useful to the newly planted plants with shallow roots. In heavy soils the root activity of the cover crops and organic matter addition help in breaking the compact clay soils, the infiltration of irrigation water is also improved. With the improvement in soil structure, fruit tree roots can spread and penetrate deep subsequently manage to collect more water and nutrients for the growth of fruit trees. The profuse growth of the fruit trees lead to higher yields than the normal.

Cover crops provide following additional benefits :

- Soil become easier to work
- Soil do not become boggy when wet, hard and dry.
- Harbour many predator.
- The activity of earthworms is increased
- The dust in orchards is reduced
- Weeds remain under check.

There are few problems of growing cover crops in orchards.

Poor Growth

Due to shade of the orchard trees the seed germination is affected adversely. The 'wattar' conditions remain for a limited period and growth of the germinating cover crops get reduced.

Irrigation

Additional irrigations are required to meet the need of cover crops. Irrigation systems are prepared/designed to meet the water requirements of the fruit trees, not for a cover crop. In summer months fruit trees can starve of water. Frequent irrigations are needed for the cover crop for its establishment. This can not be carried out through modified basin or basin systems of irrigation.

Legume cover crops like cowpeas, peas, beans, clovers serve as a better cover crop. It adds nitrogen in proportion to the amount of top growth. Arhar can be grown in sandy areas. It will protect the soil from wind erosion.

Cereals like oats, barley, wheat, rye, sarson can be grown cover crops. The cereals may not fix nitrogen but due to their vigorous root system it is a good alternative to legumes. The cereals are less prone to insect attack. Cereals produce more dry matter than many legumes.

The growth of a cover crop should be checked before the active growth in fruit trees begins. If the growth is not curtailed the cover crop will compete for water and nutrients with the fruit trees. The crop should be either slashed down to leave the cuttings on the soil surface. Some herbicides can be sprayed. It can be mechanically ploughed/disc and the cover crop is left on the surface of soil. The prime objective of cover crops is to keep the cover on the soil. Due to economical reasons, this is rarely done in India. In north India orchardists grow inter crops in orchards.

Inter Cropping

During the pre bearing period (juvenile period) farmers usually want to supplement their income by growing some crops. Most of the growers opt for wheat and fodder crops. A few grow vegetables and very few opt for pulses and legumes. Experience has shown that if intercrops are not grown judiciously these cause great harm to the fruit plants. Mostly, farmers revert to cereal crops as the fruit trees die slowly year after year. At present most of field work is done with tractor implements. The plant trunks and branches get injured by the implements, which latter on become cause of plant death. Frequent disc ploughing is very harmful to the fruit plants. It damages the feeder root system of fruit plants. Discs also make segments of various weeds roots. The segmentation of barugrass roots and shoots of 'darya booti' give rise to increased density of these weeds year after year. Injury caused by spade hoeing of tree basins can also become fatal to the fruit plants in early stages of their growth. Although cleaning of basins of weeds and mixing of manures and fertilizers is very necessary for the health of the fruit trees, even then trunk and root injuries become dangerous. Judicious use of implements in the orchards is very necessary. Growing of crops at the cost of plant health is not advisable. Potato growing has been very harmful to the fruit plants, particularly kinnow and peaches. Fodders like berseem and jawar are equally harmful to fruit plants. Many orchards of patharnakh declined due to root rot, where barseem had been grown for a number of years. These crops should not be sown in orchards. Short duration crops and pulses are beneficial to the fruit crops and can be grown successfully in orchards repeatedly. If wheat is grown as intercrop which is usually grown through out Haryana and Punjab and in some areas in Uttar Pradesh, then a separate irrigation system for fruit plants should be prepared. This is very necessary because during April no irrigation is applied to wheat and long interval of non supply of irrigation to the fruit plants shall adversely affect the growth and development. In addition to generate income from the cereals as intercrops, these crops add some organic matter through the decay of roots and trash.

It is always beneficial in the long run not to grow intercrops in orchards. Proper care of the fruit plants will add more growth which will inturn reduce the juvenile period by one to two years depending upon the fruit crop. Withgood care kinnow comes into bearing in third year, mango in fourth year and pear and litchi in fifth year after their planting. Thus, farmers can earn more profits from fruit trees by just avoiding growing of inter crops. Fillers as fruit plants like plum and peach in mango and litchi and papaya in many tree crops can supplement more income than intercrops. Non growing of intercrops can help in the production of good quality fruits for longer years than otherwise. Pear root rot, citrus root rot and decline of peaches can only be checked by avoiding intercropping. One of the major reasons of orchard failures is growing of intercrop year after year. Growers are advised to grow intercrops judiciously only for the first two to three years after planting an orchard. Irrigtational and nutritional needs of fruit tree should be met separately as per needs of the fruit plants.

(*c*) Clean Cultivation

Some farmers prefer clean orchards through out the year than growing of legumes or intercrops. They are more interested in weed control in the orchards than the health of fruit trees. In this process constant ploughing or sprays of weedicides damage the root system. The feeder roots get killed which lead to nutrient deficiencies and root rot, finally early tree decline sets in. Clean cultivation to a limited extent can be useful. It improves the soil aeration by pulverizing the top soil. The nutrients which would otherwise have been utilized by weeds can be saved for fruit plants. Hoeing breaks the upper soil crust, which checks water loss through capillaries from the orchard soil. Clean cultivation damages soil structure and soil become devoid of organic matter. It is a costly practice to keep the orchards free of weeds, through out the year. Regular cultivation destroys feeder rot system of the fruit trees from the top 10cm of soil. In most of the orchard soils the top layer is very fertile and favorable layer for feeder root system to flourish. Some stionic combinations spread most of the root system in the upper 10-20cm of the soil for example pear on *Pyrus calleryana* (pear root suckers) and citrus on Cleopatra. All own rooted layers and cutting of plum, litchi and guava plants have developed feeder root system in the upper layer. Allowing root growth in the upper zone is particularly beneficial in soils where lime induced iron chlorosis is a problem.

Some orchardist feel that frequent deep cultivation will increase water infiltration to the deeper layer of soil to be used by the fruit plants, in the long term it is detrimental. The loose surface soil created by frequent cultivation shall form a solid impermeable layer when irrigated. Soil microbes which continually work to break down organic matter, shall accelerate the break down due to cultivation, leading to decline in organic matter level. Frequent cultivation is a short term remedy for improving water infiltration. The only answer for water conservation is to grow green manuring and cover crops and addition of gypsum.

Excessive cultivation in peach, plum and kinnow orchards has resulted in to early decline. It also results into root suckering on damaged roots particularly in pear plantations propagated on pear root suckers (*Pyrus calleryana*). These root suckers become a nuisance and are difficult to control once these start appearing. Accordingly the frequent cultivation of orchard soil is not recommended.

Mulching

This is one of the important system of orchard soil and moisture management. Any material used to cover the surface soil to reduce evaporation or to check weed seed germination and suppress the vegetative growth of weeds is called mulch. Some of the important materials commonly used as mulch are sawdust, hay, weeds, straw or cut grass, leaves and stubble (crop residues). The dust mulch created by repeated ploughings or working of a rotavator to pulverize the upper soil layer also creates a barrier to soil evaporation. Mulches are capable of reducing evaporation from soil surface and are being successfully used to increase productivity and quality of the produce particularly in strawberries and fruit trees which require frequent irrigations *e.g.* litchi, peach and papaya etc. The use of moisture saving synthetic materials such as plastics in the nursery production is widely practiced. The plastic of desired width is spread in rows and fastened to keep intact to the soil in the nursery. In areas where rainfall is scanty a combination of cultivation and mulching is very useful in conservation of soil moisture. Mulching controls both, weed seed germination and weed growth. Plastic mulching is also being employed in the production of fruits like pineapple, strawberries, papaya and pomegranates. The cost of plastic to cover the surface soil limits its use in other trees fruits.

It has been seen that black plastic absorb more radiation and releases very little. The mulch becomes hot and heat is not transferred down word by conduction due to formation of intermediate layer of still air. On the other hand translucent plastic transmits much of the radiation to the soil. Hay mulch absorbs radiation just like dust mulch, but transmit very little. The hay gets hot but soil below remain cool. Gravel mulch suppressed more evaporation than dust mulch. The effect of plastic will decrease due to increased shade of the foliage of fruit trees. The crop which leave a high percentage of the residues on the soil surface could be adopted in orchards for use as mulch. The crop residue can be spread uniformly on the soil surface. The vacant spaces can be tilled with the use of rotavator, which will allow the crop residue to mix with upper 2-3cm of soil. It has been noticed that mulching with crop residue can invite attack from white ants in the orchards. The danger of fire is also there in areas with low rainfall. The mulching of plants basins with available local material such as rice trash, sugarcane trash or banana leaves etc., is the cheapest method of mulching.

2

WATER MANAGEMENT IN ORCHARDS

Importance

Availability of water for irrigation purposes is limited but demand is increasing day by day. The land which remained uncultivated for years in the past have been brought under cultivation recently. In ninety sixties only one crop was taken per year from most of the lands and there after fields were kept fallow. Now, due to increasing pressure of the growing population the intensity of cropping has increased. To raise more than two crops a year from a field, water for meeting irrigation needs of these crops is required. Fifty years earlier water table in central Punjab was about 10 meters deep, which has gone down to thirty meters. In most of the areas submersible tube wells have been established to lift water as deep as 100 to 250 meters. Under these circumstances water has become the most costly commodity. Each passing day water is becoming a limited factor for crop production. While irrigating orchard plantations care have to be taken, as orchards can not be treated as cereal crops. Most of the citrus and peach orchards declined at an early age and pear plantations got infected with root rot, due to faulty method of supplying irrigation *i.e.,* through flooding one to ten acres without segments. Judicious use of available water for raising fruit plants is need of hour. Orchards can be irrigated through several techniques. The technique may vary depending upon the topography, soil type, kind of fruit plant and source of available water. Some standard methods are flooding, sprinklers and drip. Any method can be selected depending upon the source of water and resources with the growers. The only aim should be high efficiency of available water use. A permanent irrigation system which can reduce the cost of irrigation and loss of water through percolation and run off shall be helpful to produce successful orchards on long terms. Tube wells with underground irrigation system can be a good option. The expenditure being incurred on the cleaning of water channels shall also be saved. At present 99% of the orchards in north India are flood irrigated as a single segment of one to ten acres.

The power supply to the tube wells is usually during nights, hence for convenience growers switch on the tube wells in the evenings and orchards get irrigated automatically. No labour is employed to manage water. The water flows from one end to the other end of the orchard on the basis of the slope. This practice leads to wastage of very heavy amount of water through percolation. Through flooding water may stagnate or stand up to 10cm deep at some place and there may be no water at the other. The standing water can cause root rot and leaching of costly nutrients. Some important physiological processes get adversely affected leading to leaf chlorosis, gummosis and bark scaling, root rot and finally the plant death. Application of saline water should be avoided. However, saline water can be applied by mixing with canal water. The presence of elements in excess in the under ground water such as iron, selenium or arsenic etc. can create imbalance in the nutrient uptake. These elements develop toxicity in the plant system. The water may be got tested in a testing laboratory and suitable measures adopted.

Functions of Water in the Plant System

Water performs many physiological functions in the plant system. These are as follows :

- Nutrients present in soil solution are transported from soil solution through roots to the leaves by water movement.
- Water regulates the temperature for enzymatic activity in the various plant tissues.

- It maintains the turgidity in the plants.
- Water is a major constituent of various plant tissues.
- Water molecule act as a source of H^+ for photosynthesis reactions.
- Carbohydrates, proteins and other cyclic compounds get denatured in the presence of water.

A small percentage of water is actually utilized by the plant to perform various functions out of total water absorbed from the soil. Rest of the water acts as a balancing factor for plant temperature. The purpose of irrigation is not only water supply, but also include saving of water. Hence, the method of irrigation may vary from place to place depending upon the age of plant and source of irrigation. To calculate water requirement of fruit plant a number of factors are taken into consideration. Water use efficiency varies from one kind of fruit plant to the other kind and species to species. Newly planted plants may require water frequently than the grown ups of the same kind. The transpiration ratio with dry matter produced can be calculated in different cultivars of a fruit plant and water supply adjusted accordingly. However, it is a tedious job to calculate the water requirement for individual cultivars. The use of tensiometers shall be the most appropriate scientific method to know irrigation requirement of the orchard in different climatic zones. Scientists and orchardist working with irrigation needs of orchards may need to know about soil moisture present in soil at different tensions.

Soil Moisture

Soils hold water in its pores through the process of adhesion and cohesion . During rains or flood irrigation the water in excess, percolates down is called **Gravitational water.** This water is not used by the fruit plants. The water which moves upwards from the deeper layers of soils through capillary space of the soil particles is called **capillary water.** When sufficient water has lost through capillaries. Some water is still left in the soil particles, which is beyond the reach of plant roots. This water is tightly held by the soil particles and can not be absorbed by then plant roots. This water is known as **Hygroscopic water.**

Field Capacity

Water which is held by a soil when macropores have drained the water and micropores are still filled with water, the soil is said to be at field capacity. The matrix potential at field capacity generally ranges between -0.1 to -3.3 bar. At field capacity plants remove water easily from the soil. The lowest limit of field capacity when little water is left in soil is termed as **Wilting point.** Plant may show temporary wilting at mid day during summers, which can be recovered by the fruit plant if irrigation is applied immediately. But, when plant show permanent wilting of foliage and these can not be revived by application of irrigation, the soil moisture is known as **permanent wilting point**. The soil water potential at this stage shall be - 15 bars. The fruit tress can not be revived once the soil moisture has reached **permanent wilting point**. The water supply can be regulated in the orchards by using various techniques.

MEASUREMENT OF SOIL MOISTURE AND TENSION

Soil moisture can be measured directly or by observing soil moisture potential (tension).

1. Gravimetric Method

It is the measurement of soil moisture content in a given mass of soil. Moist soil sample is taken from the orchard and weighed. The sample is dried at 100°C in an oven for 24 hours. The dry soil sample is again weighed. The difference of weights represents the soil moisture in the sample. It can be expressed on mass or volume basis. Normally water supplied to orchard is measured on acre inch or cm. This is the amount of water required to irrigate one acre of an orchard to a depth of four inches or ten centimeters.

2. Measurement of Resistance

Porous material as gypsum, nylon or fiberglass is used to measure electrical resistance. Gypsum blocks with electrodes are placed in the moist soil. These blocks absorb water from the soil in proportion to the moisture content of the soil. The electrical resistance of these blocks is related to the amount of water absorbed by the blocks. The calibration is done in the laboratory, after this these blocks are placed in orchard soil from where blocks absorb water. Electrical resistance is noted when equilibrium has reached. This electrical resistance measures the amount of water absorbed by porous blocks in turn of water content of the orchard soil. These blocks measures content in the range of -1 to -15 bars potential.

3. Neutron Scattering

The method is based on the ability of hydrogen atoms in water to check the speed of fast moving neutron and scattering them. There is one draw back in this method. That is most of the hydrogen in soil is present in organic compounds in addition to water.

Working of a Neutron Moisture Meter

The aluminium access tube having a source fast neutron and a slow neutron tube placed into the soil in the plant root zone.

Neutrons emitted by the source at a high speed collide with atom of hydrogen in soil water. The direction of movement of neutrons is changed and some energy is lost. The detector tube and scale measures the slowed down neutrons. This reading on the scale relates to the soil moisture content of the given orchard soil.

4. Tensiometer

The tensiometer determine the moisture tension/stress in the field. Tensiometer have a porous ceramic cup at the end of the tube and through which water can move out to the soil in vicinity. On the other end of the tube a vaccum gauge is fitted at the top. Water filled tube reads zero in vaccum gauge when cup is placed in soil depth with low moisture content, the water from porous cup is drawn causing a decline in water pressure in the tube till equilibrium is reached. The gauge measure this pressure called tension. The length of the tube varies in different tensiometers. The porous bulb should reach the depth of the soil where most of the fruit tree roots are present and draw water. Tensiometers are successfully used to irrigate orchards. Moisture requirements for different fruit plants have been worked out and growers can easily read the soil moisture tension of his orchard at different locations and depths with the help of a tensiometer and can adjust frequency of irrigation accordingly.

Methods of Irrigation

Different methods are being adopted to irrigate orchards in different parts of the country, depending upon the source of irrigation and agro-climatic conditions prevailing in the area. Some methods are discussed below.

1. Free Flooding

In this method no importance is given to water economy or efficiency of irrigation system, rather sole purpose of the orchardist lies in labour economy. For orchards this is the least desirable irrigation method. Entire area is irrigated. Water flows from one direction to the other end due to minor slope of the field. Most of the water gets lost by gravity. The root system of fruit trees can get damaged due to excess of irrigation water. Later on various fungi can attack the roots. These fungi are always present in the soil and only need adequate environment to flourish. Although this method is still widely used particularly for cereal crops is also being used for orchard plantations. For successful orcharding flooding should be avoided in young plantations.

It is the cheapest method of irrigation. Interval of irrigation depends upon the turn in canal irrigation system. It has many disadvantages *viz*. There is over use and wastage of irrigation water. Costly nutrients get leached. Roots remain submerged for a long time and may rot. Rotting attract fungi which cause extensive damage to the root system, may lead to plant death. Since whole area is irrigated, weeds flourish and become a nuisance.

2. Channel/Furrow Irrigation

In this method two long parallel ridges one on either side of the plant (at least 50 cm away from the plant) are prepared. These channels are connected to main supply channel, which is directly connected to the water source. The water flows from one end of the channel/furrow to the other as in free flooding. Several channels can get water simultaneously depending on the capacity of source of irrigation. The water supply in the channel is disconnected from the main channel when still two to three plants are yet to receive water. These plants automatically get irrigated with the flow of water in the channel. Thus a thin layer of water is applied to newly planted fruit plants. There is no fear of stagnation of water near the plants, there is no chance of damage to the plant root. It is very economical method of water supply. Labour economy is also there as ridges are mechanically made. The same furrow/strip can be used for two to three years, depending upon the nature of the fruit plant. The only disadvantage is that manures and fertilizer can get accumulated with the ending fruit plants. This method is a modification of the basin system of irrigation.

3. Basin System

In this method irrigation is applied just like in channel system, but the channel linking the plant basin (rings) is of small size. Basins are made around plants with a radius of 50cm in first year and then increasing to 1.0 meter and so on in the coming years as the plant canopy grows on. This increase in the basin size is necessary to receive more water for the growing need of the plant. In this method also manures and fertilizer get accumulated in last few plants.

4. Modified Basin System

This is the most practiced and beneficial method of orchard irrigation. This method is useful in level fields. Here the channel is prepared in between two plant rows. Small channels are made to connect the circular basins around the plants as in basin system, with the central water channel. The radius of the plant basins is increased at regular intervals to meet growing need of the fruit plants. This system has many advantages. It is less costly and only desired quantity of water is applied. Nutrients from one plants to other do not move. Weeds can easily be controlled by ploughing, as alternate rows do not have any channel. Disadvantage is the more cost involved in the preparation of circular basins manually, and to clean the central channels. Labour is also necessary to check and apply irrigation to individual basin.

5. Spot Irrigation

It is one of the most economical method to irrigate individual plants on uneven and plain lands or both. Water is applied in a ring around the root zone through a flexible pipe, tanker or bucket depending upon the distance of water source. It is a laborious method. It is useful in places where water is scarce and land is uneven. Mulching can be advantageous in such a situation.

6. Drip Irrigation

As the name shows, the water is supplied drop by drop in the root zone/near the root zone in desired quantity as and when required. This method resembles under ground furrows. It is a very costly but very successful in hilly areas and in light soil or sloppy lands in the plains. This method has not been adopted on a large scale in India. Due to poor after service provided by the companies, the orchardist revert to other irrigation methods again. This method has several advantages *viz.* very economical use of water. Evaporation and erosion loss of water is nil. Labour required is bare minimum. Availability of nutrients to plants is very high since, moisture in the root

METHODS OF IRRIGATION

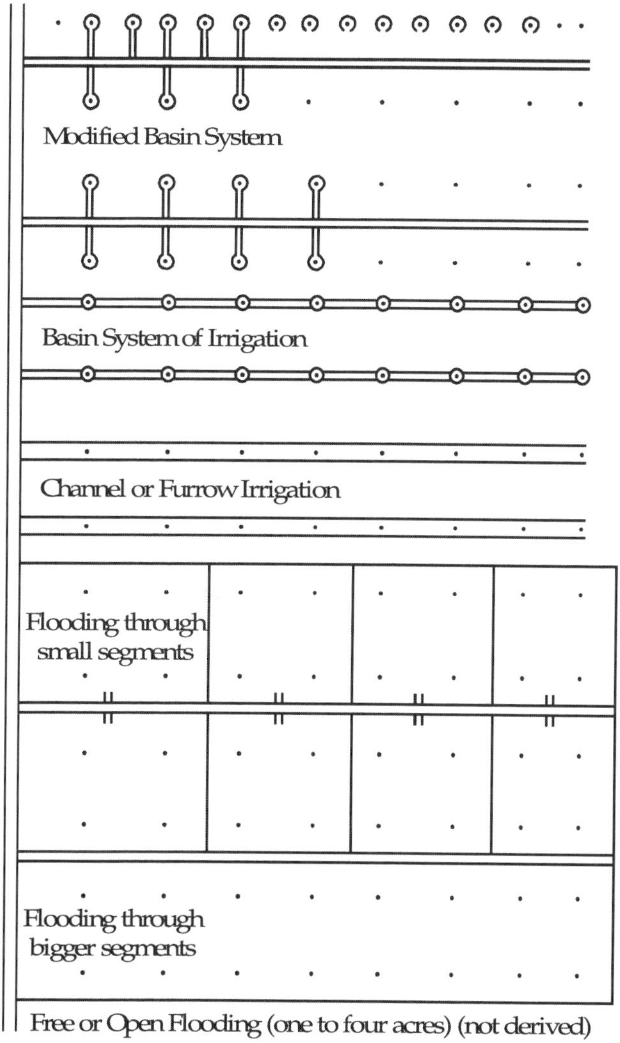

zone remains near field capacity. Fertilizers can also be applied through the system uniformly. Fungicides and insecticides can also be applied directly to the root zone, as and when required. No weeds can flourish in the vacant space due to non availability of moisture. It has few disadvantages too. It is a very costly method hence some subsidy from the government is needed. It has some technical problems too. Plant roots usually flourish in the upper strata of soil as the moisture is easily available there. During severe summers water supply may fall short of needs of grown up tree. The water pipes can be damaged by labour during hoeing.

7. Pitchers

This is an indigenous method of supplying water in the plant root zone. The pitchers of size 10 to 12 litres are placed deep in soil near the fruit plants. No hole should be made in the pitcher at the bottom. The neck of the pitcher should be kept above the soil level. The neck should be covered with the lid to avoid water loss through evaporation. The water is filled in the pitchers once or twice a week. On an average two to three litres of water is lost from the pitcher in the soil daily. This method has been very useful in filling the gaps in the orchards. The method is used on undulated lands and areas where water is scarce. The only disadvantage is that the pitchers are to be replaced after three to four years due to blockage of pores with growth of roots.

8. Sprikler Method

In this method water is sprayed under pressure. The spray is formed by the use of metal or plastic nozzles. It has some advantages like it is also useful on uneven lands. The desired amount of water can be supplied. Fertilizers and insecticides can also be applied through spray. The disadvantages are; it is a costly method and it creates humid condition in the orchard which encourage fungal and insect attack.

Interval of Irrigation

Interval of irrigation depends upon factors like type of

fruit crops, root stock used, climate, type of soil, fruiting or non fruiting and stage of fruit development. Since, major constituent of fruit is water hence, water application at development stage of the fruit is much higher than just after fruit set. During summer the requirement of the water is high hence, interval of irrigation shall be less. Normally, irrigation interval is small during early stage of plantations that is up to 2 to 3 years of age, each plant require frequent irrigation. In summer interval may be 4 to 7 days and in winter it may increase from 10 to 20 days. During dormancy of deciduous fruit plants interval may be of a month or so. During rainy season no irrigation may be applied. Canal water can be utilized by mixing with the poor tube well water. The most scientific method of determining interval of irrigation for defined fruit crop in a particular region is to find evapotranspiration rate at different times of the year. Evapotranspiration includes water loss from the soil (evaporation) and moisture loss through plants leaves (transpiration).

Rain Water Harvesting

The water table is falling very fast in north India. In Punjab 50% of the blocks have been declared as black zones due to fall of water table. Therefore, it is essential to conserve rain water, which is otherwise lost. While irrigating orchards most economical method of irrigation should be adopted. Flooding of entire area should be avoided. The rain water should not be allowed to flow out of the orchard at any cost. To get the rain water absorbed easily without standing in the fields, one many opt for digging 10 to 15 feet deep pits of 3 to 4 feet diameter, scattered in the whole orchard. Allow the excess of rain water to percolate through these pits. To avoid soil erosion into the pits, prepare a ridge around each pit 15cm of height. To check the direct flow of excess rain water into pits place one or two pipes 5cm higher than the actual soil level. This will help in percolation of water in whole orchard soil and shall limit the chance of falling earth in pits. The same pits can perform the function for a number of years.

Effect of Water Stress on the Fruit Plants

Any stress is harmful to the fruit plants. Under water stress conditions the root system of fruit plant do not develop to the desired level resulting into poor canopy growth. Prolonged water stress condition may lead to the death of feeder root system. In such a situation the leaves do not attain proper size. Prolonged moisture stress leads to premature defoliation. The older leaves get affected first. Premature defoliation, trunk and scaffold gummosis and bark scaling may result into plants decline and ultimately death. Flowering and fruit development also get affected due to little water stress at a crucial period. Fruit yield is directly affected by the water stress at fruit development stage.

Effect of Excess of Water on Fruit Plant

Excess of water in soil or in atmosphere can adversely affect the fruit plant as well as fruits. Various pathogens may attack the roots, the stem and branches and fruit due to the excess of moisture. The pear root rot, root rots of citrus and guava orchards is associated with poor drainage of soils or over irrigations through flooding. Gummosis from the trunks and scaffolds is also the result of high orchard humidity and moisture in the root zone. The attack of lichens on litchi foliage is due to high orchard humidity. Excessive orchard humidity also leads to the development green algae on tree trunks and scaffolds.

Excessive irrigation leads to more succulent growth as a result the fruit plant start bearing irregularly. Over irrigation at full-bloom stage may lead to excessive flower drop. The fruit quality is adversely affected due to excess of moisture. In kinnow the fruits become insipid in taste and fluffy after ripening. Guava, ber and pomegranate fruits show very high percentage of fruit splitting due to over irrigation or due to high orchard humidity during rains.

3

NUTRITION MANAGEMENT

Importance of Balanced Nutrition

Fruit trees require nutrition according to their ability to produce the quantity and quality of fruit; bearing behaviour whether regular or alternate; bears once in a year or more number of time in a year; type of bearing whether bears on spurs or on new growth and nature *i.e.*, evergreen or deciduous etc. fruit plants need both macro and micronutrients for the production of good quality fruits for longer period. Nutrition is also required for early completion of the juvenile period. The application of N, P and K fertilizers in split doses have given better results than one time application. The field trials on various fruits to determine the nutrient requirements in different zones in Punjab have yielded the desired results. No micronutrient deficiency has been observed in various fruit plants during their juvenile period in soils with moderate fertility level. However, it was noticed in light sandy soils particularly in foot hills and kandi areas.

The fruit plants in their juvenile phase have shown deficiency symptoms of iron, zinc and manganese. The application of N, P and K in excess doses during early years lead to fast and luxuriant growth. In the latter years, these plants over bear and show deficiency symptoms of micro-nutrients earlier than those fed with optimum nitrogenous and phosphatic fertilizers. The deficiencies were also pronounced in plants, which were fed with little or no organic manure. The orchards where green manuring had been carried out did not show any type of nutrient deficiency, rather required lesser dose of nitrogenous fertilizer. It is important to provide an optimum dose of inorganic fertilizers, which is species and cultivar specific.

The orchardist should bear in mind that organic manures as well as inorganic/synthetic fertilizers should be applied at proper time and required dose depending upon the age and bearing capacity of the fruit trees. Large amount of nutrients are removed annually from the soil by the bearing fruit trees. After irrigation, nutrition is the most important factor which will determine the success or failure of an orchard. The research work in various parts of the country on the nutritional requirements of fruit plants show wide variation in the recommendations. This may pertain to climatic conditions; soil fertility levels, rate of growth in fruit plants and yields obtained. Orchardist usually apply fertilizers as per their own experience or economic back ground. The fertilizer recommendations given by institutes are not holding good at many farmers fields this may be due to wheat rice rotation and burning of trash every year. However, these recommendations should form the base for further modification at the grower's fields.

The soil testing after 2-3 years and leaf analysis each year should from the basis of fertilization for getting high returns from the orchards. The optimum levels of nutrients in leaves of most of the fruit crops have been worked. Just as orchardist provide irrigation by looking at the leaves for wilting and adjust the irrigation interval to their crop, in the same way leaves provide adequate judgment of nutritional needs of the fruit trees. The issue of fertilizer requirements to most of the fruit trees is as debatable now as it has been forty years earlier and shall remain so in the coming times due to various reasons. Some recommendations have been made in this book in different fruits based on experiments on nutrition at Fruit Research Station Gangian of the P.A.U and at farmers field.

Sources of Various Nutrients

Form in which absorbed and source of nutrients elements.

Element	*Form*	*Source/composition%*
Nitrogen	NH_4^+, NO_3^-	Sodium nitrate 16% N; Ammonium sulphate 20.5% N; Calcium ammonium nitrate 20.5 and 25% N; Urea 46% N and Diammonium Phosphate 16 and 18% N
Phosphorus	$H_2PO_4^-$, HPO_4^2 and PO_4^3	Single superphosphate 16% P_2O_5; Double superphosphate 320%P_2O_5; Triple superphosphate 46-48% P_2O_5 and DAP 48-46% P_2O_5.
Potassium	K^+	Muriate of potash 58-60% K_2O; Potassium sulphate 48-50% K_2O; Potassium nitrate 44% K_2O
Calcium	Ca^{2+}	Gypsum 17% Ca, Dolomite and Apatite and Super phosphate 25-30% Ca
Magnesium	Mg^{2+}	Dolomite 5-20% Mg; Muscovite and Magnesium sulphate 16% Mg.
Sulphur	SO_4^2 , SO_2-	Gypsum 18.6%S; Copper Sulphate 18.8% S and Zinc Sulphate 17.8 S; Manganese Sulphate 13% S; Ferrous Sulphate 12.8 % S.
Iron	Fe^{2+}, Fe^{3+}	Ferrous Sulphate 19% Fe, Ferrous Oxide 77% Fe, Iron Chelate 5-14% Fe.
Manganese	Mn^{2+}	Manganese Sulphate 26-28% Mn; Manganese Oxide (MnO) 41-68% Mn; Manganous sulphate 53% Mn.

(contd...)

Element	*Form*	*Source/composition%*
Zinc	Zn^{2+}	Zinc Sulphate (Heptahydrate) 22% Zn; Zinc chloride 45-52% Zn; Zinc Oxysulphate 52% Zn and Zinc Chelate 8-14%Zn.
Copper	Cu^{2+}	Copper Sulphate (Monohydrate) 35% Cu; Copper Sulphate (Pentahydrate) 25% Cu; Basic Copper Sulphate 13-53% Cu; Copper chelate 13% Cu.
Boron	$B_4O_7^{2-}$	Borax 11% B; Sodium
	HBO_3^2	tetraborate 14% B; Boric acid
	H_2BO_3	17% B and Fertilizer Borate-65.20% B.
Molybdenum	MoO_4^{2-}	Sodium molybdate 39% Mo; Molybdenum Sulphide 60% Mo and Ammonium molybdate 54% Mo.
Chlorine	Cl^-	Water

Functions and Deficiency Symptoms of Essential Elements

Most of the nutrients required for plant growth and development are usually present in the soil it self. C, H, O, N, P, K, Ca, Mg and S are macronutrients because these are needed in large amounts. Out of macronutrients C, H, O is available in gaseous form and there is no dearth of these basic nutrients. Dry matter of plants contain over 95% of these elements. All the nutrients present in the soil may not be easily available for absorption by the plant roots. Several soil and environmental factors govern their availability. Some secondary nutrients (micronutrients) are also required for maintaining the growth and productivity of fruit trees/plants. These are Fe, Zn, Cu, Bo, Mo, and Cl. The functions and deficiency symptoms are briefly discussed below.

Nitrogen N

Nitrogen is a constituent of amino acids, amides, proteins, enzymes, vitamins, coenzymes and plant hormones. It imparts vigour to the plant and dark green colour to the foliage. Nitrogen is required for cell division and respiration. It delays plant maturity due to which tissues remain succulent in nature. Nitrogen also governs the utilization of phosphorus, potassium and other essential elements. It is a very mobile elements.

Deficiency Symptoms

Nitrogen deficiency causes drastic reduction in vegetative growth. It causes poor root growth and young plants give spindly appearance. Nitrogen is mobile, hence, its deficiency symptoms appear on old leaves. The leaves in the new flushes remain green. In severe deficiency young leaves also become yellow. In bearing fruit trees flowering and fruiting get reduced.

Excess Symptoms

Excess of N in plant system results in to increased foliage, broader leaves with dark green colour. The plants become susceptible to insects and diseases. The yield is low. Fruits do not develop proper colour. The quality is also not up to mark. Excess of N delays fruit maturity. It also inhibits P uptake.

Phosphorus (P)

Phosphorus is component of sugar phosphates, nucleic acids, nucleotides, coenzymes, phospholipids and phytic acid etc. It is mostly found in younger parts *viz.* flowers, maturing fruits and seeds. It enhances crop maturity, root growth, activity of rhizobia and formation of nodules in legumes. It plays a key role in reactions involving ATP and cell division. Phosphorus is required in photosynthesis and carbohydrate break down and transfer of energy within plant.

Deficiency Symptoms

Normally P deficiency has not been observed in fruit trees. It may be due to the rotation of wheat and rice in north India, where diammonium phosphate is added to both the crops. P deficiency can lead to restricted growth of roots and shoots.

The leaves become small with dull green colour, the leaves turn bronze coloured later on. Since P is mobile in plants, hence, deficiency symptoms appear on old leaves first. On acute P deficiency purple pigment may develop on the back side of the leaf lamina. In citrus fruits show puffiness.

Excess of P can inhibit the uptake of Zinc and its transport within the plant, prolonged excess can cause Copper, Manganese and Iron deficiencies. In such a situation leaf necrosis, tip die back and ultimately causes shoot death.

Potassium (K)

Potassium improves the efficiency of sugar use in plant system. It helps the plants to overcome the stresses due to environment like frost tolerance by decreasing the osmotic potential of cell sap. K regulates the supply of CO_2 by controlling opening of stomata. Young leaves, shoot tips and meristematic tissues are rich in K. It is involved in cell division. K is helpful in the fixation of nitrogen by legumes. It improves the colour, flavour and fruit size.

Deficiency Symptoms

Potassium is mobile in plants hence, deficiency appears on recently matured leaves. Acute K deficiency causes leaf burning from margins and tip. Leaves turn brown in colour and then become scorched. Shoots become thin, plants show stunted growth. In mandarins older leaves show mosaic pattern of green and yellow (speckled appearance). Fruits acquire elliptical shape.

Excess of K causes deficiencies of Mg, Mn, Zn and Fe by checking their uptake. Leaves start falling. Coarse textured fruits in citrus.

Calcium (Ca)

Calcium is present in leaves as calcium pectate. Calcium is required as a cofactor by some enzymes involved in the hydrolysis of ATP and phospholipids. It is very important for chromosome flexibility and cell division. Calcium helps in the uptake of N, Fe, Zn, Mn and Boron. Good calcium nutrition is a pre-requisite for healthy orchard particularly citrus.

Excess of calcium induces alkalinity and reduces uptake of other nutrients.

Deficiency Symptoms

Calcium deficiency causes dwarfing of plants. Twigs may die, leaves abscise. There may be a growth of multiple buds. Leaves become small and show yellowing. Fruits crack, root growth is restricted and roots may rot. Undersized and misshapen fruits in citrus is the major cause of calcium deficiency. The juice vesicles get shriveled.

Magnesium (Mg)

Magnesium is constituent of chlorophyll and protoplasm. It is involved in photosynthesis. It is an activator of many enzymes. Mg is required non-specifically by large number of enzymes involved in phosphate transfer. It is essential for formation of carbohydrates, fats and vitamins. It also stimulates phosphorus uptake and transport.

Deficiency Symptoms

Magnesium is very mobile, hence the deficiency symptoms first appear in older leaves. The most common symptom is yellowing of leaf margins and then lamina, commonly referred as bronzing or copper leaf. The yellowing starts from the base along the mid ribs of mature leaves. The yellow area enlarges and unite, only the tip and the base of the leaf remain green, showing an inverted v-shaped wedge area pointed on the midrib.

Extreme Mg-deficiency cause the plants may become defoliated completely, while limbs with little or no fruit may not show any deficiency symptom. This is particularly so in citrus.

Excess of Mg may result in K or Ca deficiency in plants.

Sulphur (S)

Sulphur is present in sufficient quantity in leaves. It is a component of cystine, methionine, proteins and fatty acids. Sulphur is also a constituent of lipoic acid coenzymes-A, thiamine. pyrophosphate, glutathione, biotin, adenosine-5′-phosphosulphate and 3′ Phosphoadenosine-5'-phosphosulphate. Sulphur retards protein synthesis, imparts hardness and vigour to the plants.

Deficiency Symptoms

Sulphur is slightly mobile in the plant system. It is bit difficult to find Sulphur-deficient orchards in north India. Some common symptoms are; the symptoms appear on new leaves. Plants remain stunted and pale green to yellow in colour. Sulphur does not move readily from old to new leaves like N. Leaf area and fruiting is reduced. Leaves fall early. The leaf chlorosis exhibited by S-deficiency usually resemble nitrogen deficiency.

Iron (Fe)

Among the micro-nutrients iron is abundantly present in soils. Iron is a constituent of cytochromes and non-haeme iron proteins. It acts as a catalyst in the formation of chlorophyll and co-factor of several enzymes. It helps in various reactions of respiration, photosynthesis and reduction of nitrates and sulphates. It has a role in N_2-fixation also. The improved Fe-polyflavonoid activity, enhances the biosynthesis of pigments like xanthophylls and carotenoids.

Deficiency Symptoms

Iron deficiency causes chlorosis in terminal leaves. In severe cases the fine network of veins is distinctively green and lamina become yellow. Twig die back due to Fe deficiency is most common in citrus. Acute deficiency causes smalling of leaves and leaf abscission. The Fe-deficiency is difficult to distinguish from Mn-deficiency.

Excess of iron has rarely been noticed in the fruit plants, this may be due to less solubility of Fe in the soil solution. However, some toxicity has been found in citrus where intervenial areas shows yellow lesions deposited on the leaf surface.

High P, Mn, Cu or Zn in soil can cause Iron deficiency.

Manganese (Mn)

Manganese accumulate in leaves more than in seeds. It is required for photosynthetic evolution of O_2, nitrogen metabolism, chlorophyll synthesis and breakdown. It is required for the activity of some dehydrogenase, decarboxylase, kinase, oxidase, peroxidase, and non-specifically by other divalent cation activated enzymes. It is also required for ascorbic acid synthesis. Mn is involved in the production of amino acids and proteins.

Deficiency Symptoms

Deficiency symptoms appear soon after the leaf is fully expanded and persists for long. Young leaves become chlorotic just like in Zn deficiency, with the only difference that there is no rosetting. The younger leaves show mottled chlorosis with green veins and mesophyll tissue yellow or white, this may spread to old leaves. Mn deficiency also resembles Fe deficiency, with only difference that the area near the veins remain green in Mn deficiency. With severe deficiency, develops light green to dull pale green splotches between main lateral veins.

Excess of Mn in soil is due to more solubility of Mn in acidic conditions. Use of acid forming fertilizers and regular sprays of $MnSO_4$ for many years can cause Mn toxicity. The appearance of brown spots on leaves coupled with growth retardation and root decay and finally leaf abscission is the result of Mn toxicity.

Copper (Cu)

Copper is associated with the mechanical strength to cell wall. It is required in oxidation-reduction reactions, photosynthesis, respiration, carbohydrate/nitrogen balance, chlorophyll and vitamin A formation, biosynthesis and ethylene activity in ripening of fruits. Copper proteins have been found in lignification, anaerobic metabolism, cellular defence mechanism and hormonal metabolism.

Deficiency Symptoms

Under field condition it is very difficult to say that the plant is deficient in Cu. Copper deficiency can only be detected through biochemical means. The total copper of the leaves may not limit for Cu deficiency. Deficiency is more pronounced with high proportion of iron and manganese. The reduced growth and a change in colour of the foliage. Twigs become angular in shape and S-shaped more or less drooping with dark green foliage. In severe Cu deficiency the tips of new shoots look shriveled and finally shed off, followed by sprouting of lower buds. The plants show a bushy growth. Dieback, gum pockets at nodes of twigs and brown excretions on fruit are common copper deficiency symptoms. Fruits have thick peel, lack juice and have insipid taste and rind cracking. In severe deficiency twigs remain covered with reddish brown droplets of gum. Fruits may split at the blossom end particularly in citrus fruits. High soil pH normally cause Cu deficiency in fruit plants on these soils.

Copper excess cause reduced plant and root growth, with less branching. Rootlets become more thick and abnormal in shape.

Zinc (Zn)

Zinc is required for the synthesis of tryptophan which is a precursor of auxin, Indole acetic acid (IAA). It regulates the equilibrium between Carbon dioxide, water and carbonic acid. Zinc is essential for carbohydrate and phosphorus metabolism and synthesis of proteins. It is a constituent of alcohol dehydrogenase, glutamic dehydrogenase, lactic dehydrogenase, carbonic anhydrase, alkaline phosphatase, carboxyl peptidase and other enzymes. Zinc improves integrity of the cell membrane and stabilizes sulfhydryl groups in membrane proteins involved in ion transport.

Deficiency Symptoms

A variety of symptoms appear in the foliage due to zinc deficiency. Its deficiency causes shortening of internodes, produces small and narrow leaves. A typical intervenial chlorosis in terminal young leaves. Acute deficiency causes rossetting terminal leaves are arranged in whorls. Whitish chlorotic streaks appear between veins in older leaves and whitening of upper leaves, characterized by irregular green bands along the midrib and main vein on a back ground of light yellow to almost white. The leaves become pointed, narrow and stand upright. The die-back of twigs starts. If the deficiency is not checked trees become filled with deadwood. Trees become unproductive.

In citrus Zn deficiency shows irregular and chlorotic leaf spots causing mottle leaf. The area near midrib and lateral veins remain green, the rest of the area become very pale yellow. Root growth get restricted. The symptom may disappear as the season advances. High levels *P*, *Ca* and *Mg* in soil induces zinc deficiency in plants.

Excessive additions of zinc fertilizers or zinc sprays can cause toxicity, which can be cured by addition of lime or by applying super phosphate to soil.

Molybdenum (Mo)

Molybdenum plays an important role in nitrogen metabolism. It is a constituent of nitrate reductase and xanthine oxidase. Mo assists in the formation of proteins, starch, amino acids and vitamins. It helps in the fixation of atmospheric N in legumes.

Deficiency Symptoms

The leaf blade may fail to expand in the growing leaves. Yellow spots develop on the lamina and gum on the lower leaf surface which turns black. Large intervenial chlorotic spots appear on mature leaves. The severely affected leaves may fall and tree may become completely defoliated. In citrus fruits large spots similar to those caused by sun-burn appear. The yellow patches on leaves coalesce into larger areas, extending all along the leaf margin, leaving the central portion yellowish green. No toxicity of Mo have been reported in fruit trees.

Boron (B)

It is immobile in plant system. It plays a role in flowering, pollen germination, pollen tube growth and fruiting. It helps in the translocation of sugars from leaves to enhance photosynthesis. Boron also acts as a catalyst in physiological processes *viz*, cell division, differentiation and development.

Deficiency Symptoms

The terminal buds fail to sprout and twigs show die back. Terminal leaves turn necrotic, shed prematurely-rosetting and apical meristems turn black. Leaves are dark green, boat like, brittle and fall early. Fruits show gummy granules in the fruit albedo with hard fruits. Seeds fail to develop with deposits around the fruit axis. The skin of citrus the fruit become hard. In some citrus cultivars fruits may crack.

Boron toxicity in lemon and grapefruits have been reported. In lemons the leaf tips burn and base of this burn is at right angle to the midrib. In grapefruit scattered yellow spots on the upper leaf surface and gummy spots on the lower surface and edge or tip burn is seen. Premature wilting in trees occur inspite of sufficient moisture in the soil.

Chlorine (Cl)

Chlorine is involved in the evolution of O_2 in primary reactions of photosynthesis, cell multiplication and turgor production in guard cells.

Deficiency Symptoms

Chlorine deficient plants show chlorosis, necrosis and bronze discolouration of leaves. Plants show wilting.

Excess chlorine results in to stunted growth, burning of tips of leaves and leaf abscission.

Optimum level of nutrients in fruit tree leaves (Based on healthy orchard observations).

Nutrient/Fruit	*Citrus*	*Litchi*	*Mango*	*Pear*	*Peach*	*Plum*	*Guava*
Nitrogen (%)	2.0-2.5	1.2-1.4	1.14-1.50	1.8-2.0	2.0-2.5	2.0-2.5	1.41-1.65
Phosphorus (%)	0.13-0.16	0.12-0.14	0.12-0.16	0.18-0.20	0.12-0.20	0.10-0.20	0.10-0.17
Potassium (%)	1.0-1.5	1.0-1.3	0.33-0.44	1.00-1.80	1.5-2.0	1.5-2.0	0.57-0.97
Calcium (%)	2.5-4.5	8-12	2.6-3.40	0.8-1.5	1.0-1.5	1.0-1.3	1.16-2.12
Magnesium (%)	0.3-0.5	0.4-0.6	0.26-0.37	0.24-0.40	0.20-0.25	0.20-025	0.31-0.51
Sulphur (%)	0.2-0.3	0.25-0.40	0.14-0.19	–	–	–	0.18-0.28
Iron (ppm)	100-150	250-300	90-100	60-100	40-50	50-60	105-153
Manganese (ppm)	25-100	50-80	75-90	25-50	20-30	30-45	58-110
Zinc (ppm)	25-50	20-30	35-50	15-25	15-20	16-20	15-29
Copper (ppm)	7-14	6-15	12-17	3-5	4-6	3-4	6-16
Boron (ppm)	30-80	26-60	–	–	–	–	–
Molybdenum (ppm)	0.1-0.2	0.1-0.2	–	–	–	–	–

DIAGNOSIS TO DETERMINE NUTRITIONAL REQUIREMENT

Importance of Leaf Analysis

Various techniques like visual symptomology, leaf analysis, soil testing etc. are being used to determine the nutritional requirements of fruit trees in different fruit growing zones. It is very difficult even today to calculate the requirement of fertilizers/micronutrients in different fruits at different ages and productivity levels. Leaf analysis have several advantages over other methods.

1. The leaf analysis gives an exact picture of the nutrient content at given time, from which a relationship with production can be drawn.
2. All the essential nutrients act in definite equilibrium in a fruit. Individual nutrient's presence or deficiency may not be related to the total quantity of that particular nutrient in the fruit tree.
3. Leaf analysis help in under standing the internal functions of nutrients in a given fruit crop.
4. It confirms the deficiency detected by visual symptoms.
5. One can distinguish nutrients which have similar deficiency symptoms. Leaf analysis helps in ascertaining that applied nutrient have entered in the plant system or not. It helps in the determination of toxicity of some nutrients.

Collection of Leaf Samples

The accuracy of leaf analysis depends upon the leaf age, its position and time of sampling. Some specific guide lines for individual fruits have been worked out for different fruits.

The leaf sampling should be carried out at a time when there is little variation in humidity and temperature. Collect samples from chest height, preferably healthy leaves without any disease or insect attack. In general mature leaves 5 to 6 months old are collected unless specified for the fruit. Sample 2% trees in the orchard diagonally from one corner to the other.

Fruits	*Leaf Age*	*or Time*	*Sample Size Number of Leaves*
Pear	Fully expanded leaves from middle of shoots or spurs	May-June	30
Citrus	Fully mature leaves	4-5 months old leaves	30
Grapes	Petioles of leaves	one month after but burst	200
Mango	Leaves along with petiole	6 month old from middle of shoot	15-20
Litchi	–do–	September	100
Guava	Recently mature leaves	August to November	30
Peach	Mature leaves mid shoot	May-June	25-30
Plum	–do–	May-June	25-30
Ber	Mature leaves	August-September	40
Papaya	Mature leaf petiole	six month after planting	20

Sample Handing and Analysis

The person collecting the samples should wash his hands before sampling. Store the samples in a refrigerator immediately after collection. The leaves may be washed at the site with deionized water and dried in shade for some time. The samples may be transported to the nearest laboratory for analysis. A minimum of three samples from orchard may be sent for analysis.

Use of Fertilizers and Manures

In the north India all the three types of soils as per reaction *i.e.*, saline, alkaline and acidic soils are under fruit plants. These soils have different problems with different fruits. In acidic soils the deficiency is found for Ca, Mg and K; in alkaline N,P, Fe, Zn, and Mn and is saline soils it is of K and Ca. It will be wise to choose the desired fertilizer for a particular soil. Right choice and suitable dose preferably in split doses, which should be applied at critical growth periods is very essential for achieving maximum fertilizer use efficiency. Dose of fertilizer and times have been specifically recommended for different fruits, which should be followed. However, leaf analysis should from the basis for any change required to get maximum productivity from the orchard. Efficiency of inorganic fertilizers can be improved by mixing the fertilizer with organic manure before its application. Fertilizers should be applied in basins uniformly by leaving an area ranging from 5cm to 30cm of radius around the tree trunks, depending upon the plant age.

Foliar sprays of micro-nutrients should be made at early stages not more than 2g/litre of water without any neutralization or by adding ½ of the quantity of lime by weight of nutrient. It will be beneficial if GA_3 is added to the micro-nutrient spray @ 10 to 20ppm. No spray may be given after the leaves have attained the age of 5-6 months. At this stage a waxy layer can develop on the leaves which restricts the absorption of the nutrient. Micro-nutrient spray should not follow insecticidal spray. There should be a difference of at lease 7-10 days between the sprays.

Manures

A decomposed mixture of dung and urine of farm animals along with roughages or the compost prepared with earth worms etc. comes under organic manures. Mostly farm yard manure is being applied along with fertilizers. However, it has been seen that vermicompost application is more useful than farm yard manure. Some orchardist also use some concentrated

Nutrient Contents of Manures

	N%	*P_2O_5%*	*K_2O%*
Farm yard manure	0.4 – 1.5	0.3 – 0.9	0.3 – 1.9
Compost	0.5 – 1.0	0.4 – 0.8	0.8 – 1.2
Vermicompost	3.0	1.0	1.5
Poultry manure	1.0 – 1.8	1.4 – 1.8	0.8 – 0.9

organic materials like oil cakes, blood meal, fish meal or meat meal.

Vermicomposting

It includes the rearing of earthworms in garbage (cow dung). Vermicompost has the capacity of improving soil fertility for a long time. It is a biological decomposition process that converts organic matter to stable humus like product. Weed seeds or pathogens present in the raw cow dung get destroyed due to the heat generated during the process. The weight get reduced so it is easy to add to the plants than farm yard manure. Though application of fertilizers gives immediate response, but it is short lived and can cause deficiency of other nutrients. The fertilizers get leached or can form metal complexes, which affect the soil pH. Due to continuous use of N, P and K fertilizers the soil texture and physico-chemical properties of soil are adversely affected. Urea which is most commonly applied to fruit crops as N fertilizer produces acids which reduce the soil pH. To avoid the use of fertilizers the growers are advised to prepare their own vermicompost which is the cheapest method of organic fertilizer production.

Vermicompost is a fine granular organic matter when mixed with soil create aeration. The mucus and earth worm casts have several growth promoters, enzymes, beneficial bacteria and mycorrhizae. It increases total microbial population of N fixing bacteria. Earthworm casts harbour vesicular arbuscular mycorrhizal (VAM) propagules which survive for over 10 months. P and N availability is increased. Quick response can be obtained as compared to farm yard manure. The quantity may be had halved than the recommended quantity of FYM.

4

WEED MANAGEMENT

What is a Weed?

Any plant growing out of its proper place or where it is not wanted/desired is a weed. Or weeds are the plants out of place in cultivated fields, lawns or orchards etc. In orchards some forestry plants like shahtoot (*Marus alba* L.); peepal (*Ficus religiosa*) and tahli (*Dalbergia sissoo*) are often seen growing. This is so because after fruit harvest and during juvenile period, birds rest on fruit plants without any disturbance. The tree weed seeds are often defecated under fruit tree. Tree plants like peepal and mulberry grow under the canopy or near the tree trunks which are difficult to be uprooted. These unwanted trees become a major problem in the orchards, as their roots and shoots get intermingled with the fruit tree both from roots and shoots. Some of the root portions usually are left in soil after their removal with spade. These roots re-sprout very fast giving out more number of plants. Weeds thus can be defined as undesirable and unwanted plants/ trees in the orchards.

Disadvantages

1. Weeds reduce the economic yield of marketable fruits by competing with fruit plants for water, nutrition, light, space and air, which are meant for the fruit plants.
2. Weeds create difficulty while carrying out orchard operation like hoeing, irrigation and harvesting of fruits.
3. The weeds are very competitive in nature and are better adapted to the areas, where they grow, example, baru

grass (*Sorghum halepense*), dib grass (*Typha latifolia*) and motha (*Cyprus rotundus*) and are capable of suppressing the growth of fruit plants in early years of planting.

4. Some climbers as weeds like chhibber (*Cucumis trigonus*) spread on the canopy of fruit plants and check the light penetration completely in the fruit plants. Most of the newly planted fruit plants get killed due to the over powering of Karari (*Convolvulus arvensis*) and chhibber (*Cucumis trigonus*) climbers.
5. Some weeds due to their rhizomes in soil like baru grass (*Sorghum halepense*) and knots of motha (*Cyprus rotundus*) resist their eradication. Weeds like *Itsit* (*Trianthema portulacastrum*) and chulai (*Amaranthus viridis*) have very high rate of seed production. The term one year seeding and seven years weeding holds true in such weeds.
6. Weeds play as alternate hosts to many serious pests/ diseases. The harbour insects/pests *e.g.*, *Cynodon dactylon* (Khabal grass) and *Cyprus rotundus* (Motha) are alternate hosts of grass hopper, like wise the 'wild senji' and 'maina' are hosts of aphids. *Parthenium hysterophorus* (congress grass) act as alternate host for mealy bugs, which attack mango, guava, pear and many other fruit trees.
7. Some weeds are injurious to animal and human health. Parthenium and Dhatura seeds are poisonous in nature.

Advantages

Although weeds have many disadvantages and are usually eradicated at a cost from orchards. At the same time weeds do have some advantages too.

1. Weeds are used as fodder for animals and some weeds are used as food for mankind. For example Chulai and Bathu are used to prepare 'Saag' and 'chhiber' fruits are used to make chatni and pickle.
2. Weeds are used to check soil erosion: In orchards where the slope/gradient of soil is more, as in foot hills, weed

like (*Cynodon dactylon*) (khabbal grass) checks soil erosion due to its mat type of growth its fine roots help in fast penetration of rain water by making the soil more porus. In arid irrigated areas, wind erosion is common and can be reduced by growing kahi (*Saccharum spontaneum*).

3. Some weeds have medicinal value for example homeopathic medicine for parthenium allergy is prepared from parthenium it self. Datura is used for animal diseases. Dried plants of *Fumaria parviflora* (Pitpapra) are used for the purification of blood to cure skin disease. Whole plant extract of *Phyllanthus niguri* (Bhoomi aonla) is beneficial in curing Jaundice.
4. Some weeds are used as breeding material for sugarcane.
5. *Improvement* of soil fertility : Weeds particularly legumes like maina (*Medicago denticulata*) and wild senji (*Melilotus parviflora*) are helpful in improving the soil nitrogen by the addition of root nodules in the soil.
6. Weeds are used in cottage industry. Leaves of *Typha latifolia* (Dib) are used for making furniture and baskets.
7. Some weeds are very sensitive to some toxins and can be used as pollution indicators for example bathu (*Chenopodium album*) is very sensitive to H_2S_2 and SO_3. Water hyacinth (*Eshornia*) can be used to reclaim industrial effluents, it can absorb heavy metals such as mercury, lead, iron, copper and tin etc.

Weeds Types

In north Indian plains both temperate and sub-tropical fruits are being grown. Even if soils in this part of the country are rich along with the availability of canal water for irrigation and sufficient rains, the orchard productivity is low due to high infestation of weeds. Weeds can be classified in different ways. Broadly, grouping of weeds can be done according to their life cycle. These can be annual, biennial and perennial weeds.

1. Annual Weeds

Weed which complete their life cycle in one season and reproduce mainly through seeds. These can further be divided into two groups *i.e.,* Kharif and Rabi weeds.

(*i*) **Kharif Annuals :** These weeds generally emerge in monsoons *i.e.,* from June to October *e.g., Cyprus rotundus* (Motha/Deela) and *Trianthema postulacastrum* (Itsit).

(*ii*) **Rabi Annuals :** Rabi weeds start growing in the end of October and set seeds before summers (April-may) *e.g., (Chenopodium album)* (Bathu), *Amaranthus viridis* (Chulai) and *Medicago denticulata* (Maina).

2. Biennial Weeds

These weeds take two seasons to complete their life cycle. In the first season vegetative growth is completed and food is stored, in the second season flowering and seeding take place. These weeds are not found in cereal crops but are in abundant in the orchards or grow along road sides or in waste lands. These weeds flourish in orchards of foot hills and at higher altitudes for example *Daucus carota* (Jangli gajjar) and *Launea nudicaulis* (Jangli gobhi).

3. Perennial Weeds

These weeds grow for two or more years in orchards or on orchard boundaries, paths and along roads. These weeds are capable of propagating through rhyzomes, stolons, root cuttings, suckers or seeds for example *Sorghum halepense* (rhizomes), *Cyperus rotundus* (knots) *Convolvulus arvensis* (stolons) and *Parthenium hysterophorus* (seed and suckers).

List of common weeds found growing in orchard of north Indian plains.

	Common Name	Botanical Name
1.	Wild Senji	*Melilotus parviflora*
2.	Jangli palak	*Rumex dentatus*

	Common Name	Botanical Name
3.	Maina	*Medicago denticulata*
4.	Pitpapra	*Fumaria parviflora*
5.	Bathu	*Chenopodium album*
6.	Karund	*Chenopodium morale*
7.	Karari	*Convolvulus arvensis*
8.	Itsit/chupati	*Trianthema portulacastrum*
9.	Dodhak	*Euphorbia hirta*
10.	Chulai	*Amaranthus virdis*
11.	Bhakhra	*Tribulus terristris*
12.	Puthkanda	*Achyranthus aspera*
13.	Chibber (climber)	*Cucumis trigonus*
14.	Amarbel	*Cuscuta sp.*
15.	Gutputna	*Xanthium strumarium*
16.	Kahi	*Saccharum spontaneum*
17.	Bhang	*Cannabis sativa*
18.	Khabbal grass	*Cynodon dactylon*
19.	Baru grass	*Sorghum halepense*
20.	Motha	*Cyperus rotundus*
21.	Parthenium or congress grass	*Parthenium hyterophorus*
22.	Choorislote	*Artemisia tridentata*
23.	Dib (Dab)	*Typha latifolia*
24.	Bhoomi aonla	*Phyllanthus niguri*
25.	Kana	*Saccharum munja*
26.	Dhatoora	*Datura stramonium*
27.	Lantana	*lantana camara*
28.	Jangli gobhi	*Launea nudicaulis*
29.	Jangli gajjar	*Daucus carota*
30.	Khatyay	*Portulaca sp.*
31.	Peepal	*Ficus religiosa*
32.	Tahli	*Dalbergia sissoo*
33.	Toot	*Morus sp.*

Method of Weed Control

Weed control methods can be classified into following :

1. Preventive Method

The best method is to prevent the entry of new weed and weed seeds in the orchards area. While purchasing nursery plants ensure that the nursery is free from weeds. No weed plant should come along the earth balls. It has been observed that new weeds come along with earth balls at the time of planting an orchard. The weeds growing in the orchard should not be allowed to flower/seed. Do not use farm yard manure which may contain weeds seeds. Clean the machinery/tractor while operating in an orchard so that weeds seeds do not enter in the orchard. Keep water channels, paths, roads and orchard boundaries clear of weeds. "One year seeding seven year weeding", clear the weeds before flowering take place. This will prevent the infestation of weeds in the orchard.

2. Eradication

Complete elimination of weeds plants, plant parts and seeds from the orchard and its vicinity, shall help in the eradication of weeds. However, it is not possible/feasible to eradicate all the weeds from any area. One may try and achieve some satisfactory result.

3. Control

The process of limiting weeds infestation to a level so that there is minimum of weeds competition through some practices, chemicals or non-chemical means which may bring down the weed population to a great extent. The weed control method may comprise physical, mechanical and chemical means.

Orchardist prefer controlling weeds by ploughing the orchard with tractor implements. This may be the easiest method to control weeds, but often the implements or body of the tractor may hit and bruise the tree trunks/scaffolds or shoots. The trunk injury can become the cause of the death of the fruit tree. The injured portion is attacked with fungi and gum may ooze out of the wound. The flow of nutrition through xylem and metabolites through phloem gets disturbed/hindered, which leads to plant decline. Complete eradication of weeds is a difficult and costly proposition. To get weeds under control to a level which does not adversely affect the health of fruit plants, a combination of methods may be adopted. Some of the measures are as follows.

1. Mechanical Method

This involves hand pulling, hoeing, tillage, mowing and smothering with non-living mulches. This is very effective method of weed control. No injury is caused to the foliage of the fruit trees. Hand pulling involved uprooting of weed manually or hoeing with khuhpa or spade. The basins can be covered with black polythene (plastic film) after hoeing, which will prevent photosynthesis in the sprouting weeds and check further growth and reproduction. Simultaneously polythene film shall check the loss of moisture from plant basins and shall help in moisture conservation.

2. Biological Method

Some insects as natural enemies, disease organisms, parasite plants and selective grazing by livestock comes under this control. Insects and disease organisms have been the most successful natural enemies of weeds. The principles involved in selection of biotic agent are :

(*i*) Biotic agent should be host specific. The agent shall not attack any plant species other than those for which it is released.

(*ii*) Its introduction and multiplication should be easy.

(*iii*) It is should be from an area that is climatically similar to the area where it is to be released.

(*iv*) Bioagents should be of different feeding habits (plant parts) so that a weed is completely killed. This will also avoid competition between two more agent released for the same food.

3. Chemical Control

A chemical use for killing or inhibiting the growth of plants is termed as herbicide. Herbicides may be classified into three groups depending upon their effect on the plant system.

(*i*) **Growth Regulators :** Growth regulators are also called growth modifiers, growth substances, translocated herbicides or systemic herbicides. The herbicides can be absorbed by root or above ground parts and are translocated through the plant system. These herbicides are species specific. The concentration at which these are effective is also species specific. This property of these chemicals make it possible to use herbicides selectively, killing certain plants without significantly injuring the others. The excess rate of application can be harmful to soil and other flora.

(*ii*) **Soil Sterilants :** Chemical which can prevent the growth of green plant when present in soil are considered as soil sterilants. Some herbicides resist chemical decomposition and leaching. On the other hand some herbicides get leached soon or can get decomposed easily. All these characters are depending upon climate, soil type and rate of use.

(*iii*) **Contact Herbicides** : These herbicides are capable of killing plants which so ever come in contact with it. There is a little or no translocation of herbicides through the system. Contact herbicides are very effective against annual weeds. These may be selective or non-selective in nature. A selective contact herbicide shall kill or suppress only those weeds without significantly injuring rest of plants. A non-selective herbicide shall kill or suppress all vegetation, which so ever come in contact with it. Living tissue has no resistance to these herbicides. A few of the herbicides are discussed below.

Atrazine and Simazine

Both these herbicides are triazines belonging to heterocyclic nitrogen compounds group. Their application inhibit Hill reaction of photosynthesis as a result leaves turn yellow. They also inhibit the non-cyclic photophosphorylation or ATP synthesis.

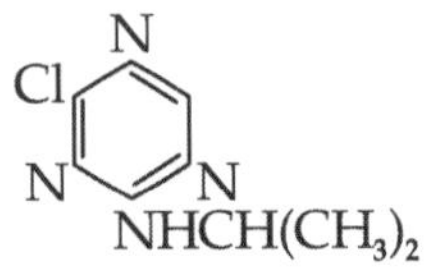

Atrazine

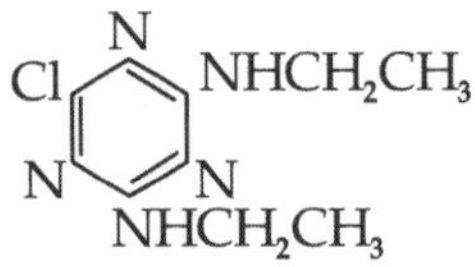

Simazine

Atrazine

(2-chloro-4-ethylamino-6-isopropyl amino-1, 3, 5- trizine). It can be sprayed as well as applied through soil application. Both roots and leaves of emerging weeds absorb the herbicide.

Simazine

(2-chloro-4,6-bis [ethyl-amino]-1, 3, 5 triazine). It is commonly used to control grasses in non cropped area. It is effective as a pre-emergence control of annual grasses and as a spray to control broad leaved weeds in citrus orchards. Citrus trees/orchards treated with Simazine or Atrazine to control weeds were found to grow better than weeded control trees.

Substituted Urea

Diuron and Isoproturon are commonly used herbicides in orchards for controlling grasses and annual broad leaved weeds.

Diuron

(3-[3, 4 dichlorophenyl]-1, 1-dimethyl urea): It is an important herbicide being used in orchards for the control of woody weeds by basal application.

Diuron

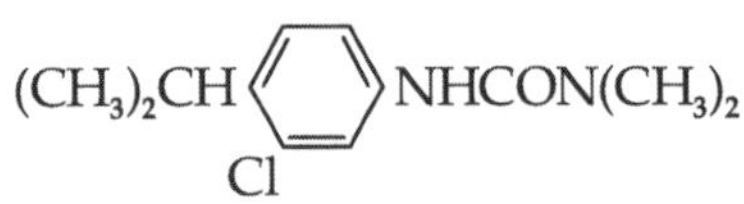

Isoproturon

Diuron, Monuron, Bromacil and Simazine are commonly used for controlling weeds in orchards. Monuron is a good herbicide for citrus orchards. Diuron @ 4kg/ ha. gives good control of weeds when used as pre emergence weedicide. Diuron @ kg/ha plus Amitrol 2.5kg/ ha. gives complete weed control in citrus orchards.

Isoproturon (3-4 (-isoproyl phenyl)-1, 1-dimethyl urea). It is an effective weedicide to killing of annual weed seedling. The seedling get killed as soon as they start photosynthesize.

Pyridines : Diquat and paraquat are widely used in orchards.

Diquat (1, 1-ethylene-2-2′ -bipyridillium ion formulated as dibromide). Diquat is usually used to control aquatic weeds. It is also used to desiccate potato haulm. Repeated applications can kill parthenium also.

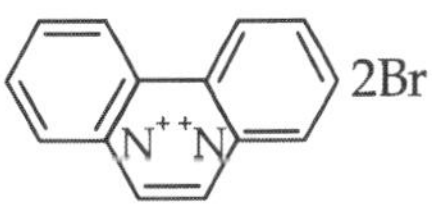

Diquat

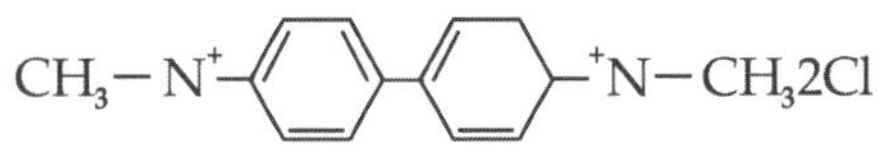

Paraquat (Grammoxone)

Paraquat (1, 1 ′-dimethyl-4,4′ -bipyridillium ion formulated as dichloride). It is very fast acting herbicide. The first effect is noticed just after few hours of spray. The plants get killed within 3-4 days. A peculiar odour is felt in the orchard after its spray. It is usually sprayed @ 6-8 ml/litre of water. It destroys the photosynthetic tissue and total green matter is killed. The photosynthetic activity is checked by removing electrons from electron transport from ferridoxin in PSI. The herbicide can be sprayed under the fruit tree canopy up to tree trunks. Repeated sprays can kill *Cyperus rotundus* (motha), *Parthenium hysterophorus* (congress grass) and *Imperata cylindrica* (quack grass/thatch grass) also.

Organophosphorus Compound

$HO_2CCH_2NHCH_2$ P = O (OH_2). Glyphosate (N-phosphonomethyl) glycine. It is a derivative aminoacid, glycine. Commonly used under the name of 'Round up ' or 'Weed off'. It is used as post-emergence herbicide. It is very effective when sprayed at well developed stage of weeds, with sufficient vegetative growth and foliage. This herbicide is rapidly absorbed by the leaves and translocated from vegetative parts to under ground parts. Glyphosate inhibits amino acid producing enzymes. Once the protein production is stopped, the growth stops. Some reports suggest that glyphosate do not degenerate in soil and percolates down as such and pollute under ground water. At the same time it is also reported that it gets inactivated on contact with the soil. What ever the case may be it provides excellent weed control in orchards if sprayed carefully. It is advisable to use this herbicide once in two years or once a year. Repeated sprays in an orchard affects the fruit trees by slowing down its growth.

Herbicides give satisfactory weed control only if these are sprayed at appropriate concentrations. The volume of water used also determine the degree of weed control in an orchard. The steady speed of walking during the spray leads to even distribution of herbicide and uniform coverage of foliage, resulting in good weed control.

FACTORS AFFECTING THE EFFICACY OF HERBICIDES

1. Soil Type and Organic Matter

The type of soil determine the dose of herbicide to be effective in the control of weeds. Soil colloids adsorb herbicides and make it non-available for action. Soils with more clay content need higher dose of herbicides in comparison to sandy soils. The soils with high organic matter content also require higher dose of herbicides than soils with low organic matter. The absorbed herbicides may be released slowly rendering the chemical non-effective or less effective.

2. Climate

The activity of herbicides is influenced by climate particularly temperature. In warm climate herbicides are more effective even at some what lesser rates than in cold regions. At the same time, a very high temperature may lead to volatilization of some of the herbicides.

Wind also has direct effect on evapotranspiration. Wind particularly during summers causes rapid drying of the spray solution on foliage resulting in reduced absorption.

Light helps in higher and fast penetration of herbicides by stimulating stomatal opening. It increases photosynthesis which enhances the movement of herbicides from leaf to other parts of the plants. When solar radiation is very high, it causes the break down of herbicides in the presence of light, which is known as photodecomposition. Herbicides prone to decomposition needs to be sprayed in the evenings and should be incorporated immediately into the soil, *e.g.*, monuron.

High humidity generally favours foliar absorption of herbicides due to increased drying time of spray droplets on leaves. The enhanced stomatal opening and phloem transport is due to high humidity. Drought conditions cause reduced absorption and translocation of herbicides.

3. Use of Fertilizers

Fertilizer use/application influence the herbicide activity. Fertilization improves herbicide efficacy by higher absorption and subsequent more translocation leading to higher metabolism of herbicides. Nitrogenous fertilizers increases the efficiency through higher absorption and translocation, where as phosphorus fertilizers increases toxicity by enhancing its metabolism.

4. Time of Application

Different weeds determine the time of application in different fruit crops. Time of application determines the depth to which the herbicides moves in the soil, plant growth and time of seed germination.

5. Irrigation

Irrigation or rain immediately after the spray can reduce the efficacy due to fast percolation of herbicide in soil.

6. Kind of Herbicide

Systemic herbicides interfere with normal functioning of one or more physiological and biochemical processes. These herbicides kill the plants by chronic toxicity. The processes inhibited are : 1. Respiration and mitochondrial activities 2. Photosynthesis 3. Phloem and neuleic acid metabolism and enzyme synthesis.

7. Herbicide Degradation

The substitution of an atom in the herbicide molecule by another will result in the formation of a totally non-phytotoxic compound. Some reactions involving degradation in plant system are; oxidation, decarboxylation, deamination, dehalogination, hydroxylation and many more.

Control of Tree Weeds in Orchards

Tree weeds create more problems than the crop weeds. It is very difficult to control tree weeds due to their extensive root system and growth. Only using herbicide spray may not be the solution. The trees like 'peepal' and 'shahtoot' should be repeatedly defoliated by spraying herbicides. The injections of killing chemicals can be given in tree trunks, by making deep cuts with an axe. The apical dominance can be broken by spraying herbicides or by heading back the shoot tips. Cut the trees from the soil level and treat the stumps with phenoxy-herbicides immediately after cutting down the tree. The stumps can also be given a higher dose of common salt (NaCl) or urea fertilizer to check the resprouting of the stumps.

5

UNFRUITFULNESS IN FRUIT CROPS

Orchardists are concerned with the profits from the fruit crop. Sometimes orchard's efficiency of bearing is low in spite of the abundant blossom. The conditions under which fruit trees do not set fruit to a desired extent is known as unfruitfulness. These conditions may be external or internal. External factors are not related to the fruit trees. Internal factors are specific to the fruit plant.

A. External Factors

The environmental conditions govern the life cycle of a fruit plant. These conditions greatly influence the flowering and fruit setting in the fruit plant. Factors like nutrition, pruning, water supply, rootstock used, temperature, insect pests and diseases affect the fruiting in fruit trees.

1. Climate

I. Temperature : This is one of the most important factor, which governs the flowering, fruiting and fruit development. Temperature requirements are species specific. Variation in day and night temperature or extreme fluctuations, or continuously hot temperature adversely affect the pollination and fruit set in most of the fruits. For example, cloudy weather at full-bloom stage affect the pollination and fruit set in soft pears and plum. Poor setting in Sapota fruit is due to drying of flowers. Temperature affects the bee activity in an orchard. Bees help in the pollination and fruit setting.

II. Rainfall : Rainfall is very important to maintain the underground level of water. It also helps in improving the quality of fruits particularly in peaches, patharnakh and mango. Rain at full bloom washes away the pollen, stigmatic fluid and keeps the pollinisers away, hence affecting the fruit set.

III. Winds : There are many fruits which get pollinated through wind (Anemophily). For these plants movement of air at the time of flowering is necessary to affect pollination. Most of the fruit plants are insect-pollinated (entomophilous). In such plants, wind hinders rather than helping in pollination.

Hot wind (loo) during May-June in North India desiccate the flowers or newly set fruits. Winds along with rain is more harmful.

IV. Frost/Freeze : It is the most important factor for deciding the fruitfulness in an orchard. Frost injury can convert a regular bearing cultivar into an irregular bearing. Even orchards may not produce any fruit for two to three years continuously due to killing of branches caused by severe frost. Even set fruits nearing maturity in Guava were spoiled by severe frost which occurred during, December 2007 and January 2008.

V. Hail Storm : Hail storm has been found to be very harmful in hilly areas. Most of the apple crop was damaged by hails at fruit set. Hails kill flower buds and blossoms. There are areas which are prone to hails and freezes every year. Some areas are almost free from these hazards in the same zone.

VI. Cloudy Weather : Cloudy weather is more dangerous than hails. The humidity makes the conditions most favorable to spread fungal diseases. Powdery mildew in mango and Umran ber usually appears in cloudy weather.

VII. Intensity of Light : Light also plays a major role in the fruitfulness of an orchard. Strawberry plants develop pistils only when these are exposed to specific light intensity. In overgrown litchi and mango plantations due to overlapping shade the fruit set is reduced. Closely planted kinnow '10 by 10' or '10 by 20' at full growth stage bear poorly due to poor light penetration. Even fruits do not develop proper color at maturity.

2. Disturbed Moisture in Soil

Soil moisture is one of the key factors to production of fruits. The excess soil moisture as well as low soil moisture conditions at the time of flowering and fruit set encourage abscission layer formation, leading flower and fruit drop. Growing cover crops or mulching the basins can overcome the low soil moisture. Flooding of the orchards at flowering time should be avoided.

3. Nutrition

When balanced nutrition is not given the plant's growth and development is affected. The practice of application of fertilizers at the time of flowering/fruit set should be avoided. The manures and fertilizers need to be applied one to two months before flowering and in split doses after fruit setting. Imbalance in the nutrients certainly cause unfruitfulness or flower drop. Higher doses of fertilizers render many kinds of fruit plants more vegetative and without flowers or produce abnormalities in the flowers. Excess of nitrogenous fertilization induces barrenness in plants.

4. Rootstocks

Rootstocks affect the scion cultivars physiologically. Quince rootstock induces dwarfing in pear due to formation of inverted bottleneck, whereas D-4 produces very vigorous pear plants. Trifoliate orange and its hybrids Troyer and Carrizo produced dwarf citrus plants than on Jatti Khatti and Kharna Khatta. The grafted plants produce early crops than those raised through seeds, this way fruiting is affected by the rootstocks. Use of interstock reduced the juvenile period by two years in 'Leconte' and other soft pears. Patharnakh (*Pyrus pyrifolia*) interstock was better than root suckers (*Pyrus calleryana*).

5. Pruning

The deciduous trees are judiciously pruned each year. Unpruned grapes bear little crop with small bunches as compared to pruned vines. Unpruned peaches bear little fruit and are prone to limb breakage. To improve fruiting, light pruning is given every year. Similarly, summer dormant ber is pruned in May-June to get more fruit. The intensity of pruning varies from cultivar to cultivar and species to species.

6. Plant Age

Some fruit plants have long juvenile period than others. These plants cannot be made to bear fruit early. The *Citrus medica* and *C. jambhiri* plants bear only male flowers in the first few years. Slowly hermaphrodite flowers appear with age. Young grape vines produce less pollen than the aged vines of the same cultivar. Very healthy plants bearless flowers than semi-healthy plants. Declining plants bear profusely than healthy trees.

7. Chilling Requirements

Some fruit trees need desired chilling hours requirement for spur formation and flowering. Bartlette and conference pears do not flower in plains; whereas low chilling requiring 'Leconte' and other pears bear heavy fruit. Apple remains vegetative for long in lower hills than higher altitudes. Southern cultivars of mango shoots get killed due to frost or low temperature in north hence no fruiting. In 'Leconte' pear profuse flowering is observed every year but fruit set is irregular due to change in season.

8. Spraying at Full-bloom

Normally insecticide sprays at flowering time are not recommended. However, sometimes due to prolonged flowering particularly in litchi, mango and pear, it becomes necessary to control aphids, etc. Hence, insecticide is sprayed, which adversely affect the fruit set. In mango a disease 'Jhumka' has surfaced due to poor pollination. Fungicides normally do not affect pollination. Insecticide sprays should be avoided on full-bloom. Spray may be advanced or delayed for a week or so.

B. Internal Factors

Some fruit species produce abundant flowers and set little fruit. Sometimes there may not be any fruit on a tree. Pollination failure, sterility or deficiency of nutrients may be the major cause of unfruitfulness. The internal factors are : (1) Impotency, (2) Incompatibility, and (3) embryo abortion. Some other factors are related to flower structure and form. These are :

1. Dicliny or Unisexuality

The stamens and carpels lie in separate flowers. Male or female flowers borne on same or different trees.

(*a*) Male and female flowers on the same plants are called monoecious, *e.g.*, walnut, pecan nut, chestnut, banana and coconut.

(*b*) **Dioecious** : The male and female flowers are borne on two different plants. Hence, to set more fruit male flowers from **male plants** are placed close to female flowers on the other plant, *e.g.* Date palm and papaya. Papaya has 8 types of flowers. (1) Pure pistillate flowering plants (2) Pure staminate (3) Both staminate and perfect flowers (4) Plants with sterile pollen (Pseudo hermaphrodite) (5) Plants producing staminate and perfect flowers but neither pollen nor pistil is fertile (Sterile hermaphrodite) (6) Plants producing staminate, pistillate and perfect flowers (7) Plants with staminate and perfect flowers (8) Plant with pistillate and perfect flowers.

2. Dichogamy

In many bisexual flowers the anthers and stigma mature on different times. This condition is known as dichogamy. It acts as a barrier to self-pollination hence; unfruitfulness in such plants is the result. When the gynoecium matures earlier than the anthers of the same flower the condition of the flower is *protogyny*. On the other hand when the anthers mature first and discharge their pollen earlier than the stigma of the same flower, the condition is *protandry*. Avocado flowers are protogynous in nature and in mango stigma is receptive for two hours but pollen is available for longer period hence such a situation is termed as protandry. Coconut can be another such example.

3. Self-sterility

In such flowers the pollen does not fertilize the ovule of the same flower through stigma; this leads to unfruitfulness in many fruits. This can also be termed as incompatibility. In incompatibility, both pollen and ovule are fertile but fail to unite due to some reason. The self-sterility has been found in pear, apple, plum, almond citrus and mango cultivars. Bartlett pear is self-sterile. Commercial cultivars of loquat are self-incompatible. In plum cultivars pollinizer Kala Amritsari have been recommended due to self-incompatibility.

4. Heterostyly

In this condition the flowers have short styles and long filaments (stamens) and other flowers on the same tree or species have long style and short stamens/filaments. This is known as *dimorphic heterostyly*. Similarly there can be trimorphic heterostyly, *i.e.*, stamens and styles of three different lengths, example pomegranate, litchi, sapota and almond. Hence, poor fruit set may be due to heterostyly.

C. Physiological Reasons

It is difficult to assign the particular reason for unfruitfulness in some of the situations. These physiological factors are;

1. Pollen Tube Growth

The rate of pollen tube growth through the style is so slow that it does not reach the ovule. It is usually so in heterostyly condition. Poor rate of pollen tube growth has been found in pear and mandarin.

2. Poor Pollen Germination

Sometimes due to physiological reasons pollen does not germinate on the stigmatic surface. This situation may bring unfruitfulness.

3. Delayed Pollination

Sometimes flowers abscise because of delay in pollination, for example, if kangji kalan lemon flowers are not pollinated from outside source, all flowers fall down due to self-incompatibility. Hence, to get good crop synchronizing in pollen cultivar with that of lemon cultivars should be included as pollinator.

4. Nutrition

When fruit plants over-bear in one season and get depleted in nutrition, there may be no crop in the coming year. Nutrition affect the pistils and pollen productivity both. For example, plums and peaches may not produce any crop in one year after a heavy crop in the previous year. Carbohydrate deficiency has been reported for coulure or blossom abortion and flower drop in grapes. Twenty percent sucrose spray at full bloom stage improves fruit set in soft pears.

6

PROPAGATION AND NURSERY MANAGEMENT

Fruit propagation is art and science of multiplication of plants. Both sexual and asexual methods are employed for the propagation of different fruit plants. Some fruit plants can be planted in the orchard as seedlings (raised from seeds). In others asexual propagation is very necessary to get uniform and precocious plant. Both vegetative and sexual (seed) methods have advantages and disadvantages.

Advantages of Raising Own Nursery

Growers do not plan in advance. They suddenly get motivated at the advice of the extension worker or a friend and start for investigating for the nursery source or lift the plants from any nursery near by. There is no private nursery in northern India, having its own well managed rootstock mother plants. Many do not have their own scion mother plants even. Private nurseries as well as some government nurseries depend upon the locals to purchase seeds of rootstocks, who collect the fruits for the extraction of seed from wild seedling trees or orchards of that area. It is disgusting that no private nurseryman is using Jatti khatti (*Citrus jambhiri*) as rootstock for citrus cultivars. The rootstock seeds include citron, kharna khatta, sweet orange and Jalandhri Khatti seeds. Ultimately, the sufferers are the orchardists. Nothing could be done at a latter stage.

Sometimes suddenly demand for a particular fruit plant is witnessed. In such a situation, growers are supplied very small-sized young nursery plants, which usually fail to establish in the field. Many gaps come up in the orchard in the first year itself. These gaps remain unfilled for many years. This affects the income from the orchard. Most of diseases, pests and weeds are transported along with the plants from such nurseries.

Some Important Rootstocks for Different Fruits

Fruit	*Rootstock*
Mango	GN2, GN5, Rumani olour and Totapari Red small.
Guava	Lucknow-49
Kinnow	Jatti Khatti and Cleopatra (Nuclear seedlings).
Blood red (Malta)	Cleopatra (nucellar seedlings)
Musambi	Pectinifera (nucellar seedlings)
Other citrus cultivars	Cleopatra and Jatti khatti
Ber	*Zizyphus mauritiana* (coimbatore) *Zizyphus numularia* (malah ber)
Sapota (Chiku)	Khirni (*Manilkara hexandra*)
Pear	Kainth (*Pyrus pashia*) and root suckers (*Pyrus calleryana*)
Peach	Khurmani and sharbat cultivars of peach (*Prunus persica*), Flardaguard
Plum	Own rooted cuttings of Kabul green gage.

It will be in the interest of the grower to plan in advance and prepare his own planting material of high quality for his orchard. This can only be done if growers plan their orchard plantation two years ahead of planting. Raise your own nursery for raising vigorous precocious and high quality plant material. Few advantages, of raising own nursery are enlisted.

1. The desired rootstock for particular cultivar can be raised for propagation.
2. Scion wood of the required cultivar from very healthy plants of known pedigree can be used.
3. The pests and diseases are kept under control.
4. The weeds free nursery material is available.
5. Well-sized plants can be planted to reduce juvenile period.

6. Gap filling can be done immediately as and when the gap occurs in the orchard.
7. The nursery plants have already been acclimatized with the soil and environment.
8. Transportation cost and mortality during lifting is nil.

Rootstock Identification in Citrus

Citrus is being propagated on many rootstocks by the different nurseries. To identify the rootstock used at a latter stage, the young leaves arising from the rootstock portion can be analyzed for isozymes. A brief method is explained for use.

Potato Starch Gel Electrophoresis

The upper two leaves from the root sprouts, which have not yet developed chlorophyll, are removed with the help of a sharp blade without injuring the stem axis. Peroxidase isozymes studies give different bands, which can be easily used to identify the rootstock. The technique used for studying isozymes of slots are made in a single line in the gel. The slots are made in the center of the plate. The freshly taken leaf samples are ground with a mortar and pestle on ice with two drops of extraction buffer. The leaf sap is absorbed on the strip of Watchman 3 mm chromatography paper 0.7*0.7 cm in size. The strips are placed on filter paper before inserting into slots already made in the gel, with the help of forceps. The gel tray is placed horizontally. The migration chamber with the gel is kept in a refrigerator to reduce the heating effect during electrophoresis. Then the power supply is turned on to adjust 250 V and current 25m A. To get clear banding pattern remove the filter paper strips after 20 minutes of run. Increase the voltage to 300 V and current to 30 m A. To complete the electrophoresis it takes about 3 hours. After this the gel is placed in a refrigerator.

Staining of Gel for Peroxidase Isozymes

The peroxidaseisozymes are stained by the following reaction :

$$2H_2O_2 \xrightarrow{\text{Peroxidase}} 2H_2O \quad O_2$$

The oxygen liberated oxidizes colorless Benzidine into a colored compound,. Bands appear immediately and staining is complete in 1-2 minutes. Prepare a sketch of the banding pattern immediately after staining and can be photographed subsequently.

Numbering of Bands

The cathodal bands carry prefix C and anodal bands carry prefix A. Symbols C and A are entirely operational as net charge of a molecule will very with pH. Usually band closest to the starting point is called 1, more rapidly moving bands are given progressively higher numbers. Since no detailed genetic analysis of isozymes is carried out, the bands can be used to identify different citrus species used as rootstock. Peroxida sesozymes bands can identify even the cultivars. The starch gels can be glycerolized and dried in an incubator at 55°C for making these permanent. The banding pattern of different citrus rootstock is given for easy understanding the difference.

Zymogram of nucellar seedling in : (1) Jattikhatti (2) Kharna-khata (3) Cleopatra (4) Rangpur lime, and (5) Pectinifera.

There can be little variation in these bands due to the zygotic origin of the same species.

Sexual Propagation

In this method, plants are raised from seeds or embryos. When the asexual method was not known, the plants were produced through this method. Even today papaya, phalsa and guava are being propagated through seeds.

Benefits

1. It is the simple and cheap method and can be employed by the growers at their own fields.
2. Plants produced are hardy and have strong root system.

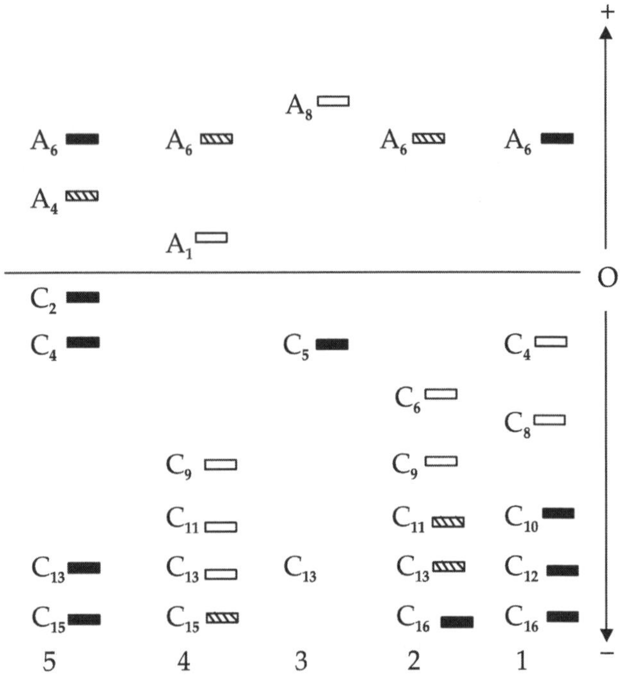

3. When the plant cannot be propagated through vegetative means, then it is the only method of propagation.

4. Many commercial cultivars of mango, *e.g.* Dusehari and Langra and Loquat *e.g.* Golden Yellow, etc. originated as chance seedlings.

5. Hybridization work depends upon seed germination.

6. Polyembryonic seeds particularly in citrus, *e.g.* Cleopatra Pectinfera and Jatti khatti have 3-5 embroys per seed. Hence, to rejuvenate the declining clone use of nucellar embryos can be made.

7. The presence of apomixis in some apple species and cleistogamy in grapes can help in the production of true to type seedlings and thus propagation through seeds.

Disadvantages

1. Plants raised through seeds have long juvenile period. It takes long time to fruiting.
2. The seedling trees have very deep root system hence the trees attain big size at maturity.
3. The bigger trees compel the growers to plant few trees/ acre. Thus in early years little yield per acre and in latter years trees get intermingled among themselves, producing poor quality of fruits due to shade.
4. Some viruses are seed borne, for example, mosaic in peaches and psorosis in citrus. These viruses perpetuate through seed.
5. Some seedless fruits cannot be propagated through seed. The only method to produce these plants will be asexual method.
6. The good affect of a rootstock particularly, dwarfness and early bearing (precocity) cannot be achieved when fruits are propagated through seeds.
7. The improved characteristics of a particular cultivar can not be perpetuated in plants raised from seeds.

Vegetative Propagation (Asexual)

The regeneration and callus formation at cuts make it easy for the propagation of plants through cuttings, layering, grafting, etc. Such method help in the production of identical plants in a short time interval at a large scale.

Advantages

1. Plants producing seedless fruits *e.g.* Marsh Seedless grapefruit, Thompson Seedless grapes, Banana, Flortasun peaches, etc. can be propagated through vegetative means.
2. Vegetatively propagated fruit plants are true to type and have less juvenile period. The tree size also remains

smaller than the seedlings. More number of plants can be accommodated per hectare leading to more income from unit area.

3. Incidence of disease and pests can be minimized by using resistant rootstocks, *e.g.* budding phylloxera susceptible *Vitis vinifera* on St. George (*Vitis rupestris*) rootstock can avoid the pest attack.

4. The difficulty of unsuitable soil could be overcome by using rootstock, *e.g.* different rootstocks perform very successfully in different areas of citrus cultivation. Chiku can be propagated on Khirni (*Manilkara hexandra*) rootstock successfully in rocky lands.

5. Some rootstocks are resistant to crease formation as Cleopatra (*Citrus reshni*) and exocortis resistance in Jatty Khatti (*Citrus jambhiri*). Trifoliate orange (*Poncirus trifoliata*) is tolerant to citrus nematode (*Tylenchulus semipenetrans*) and these rootstocks are being used in difficult situation.

6. Inferior quality of fruit trees can be top worked to improve their fruit quality on the same roots. Some times a rootstock is incompatible with a scion cultivars, then a compatible interstock can be inserted.

7. The benefit of precocity is taken by budding or grafting to see the early performance of produced hybrids.

8. Fruit quality/maturity can be improved/hastened by using a particular rootstock. Trifoliate orange and its hybrids can bring early maturity in kinnow than on kharna khatta rootstock. Quality of fruits is improved on Cleopatra rootstock.

9. Some times roots of the bearing plants can get damaged. To save these plants new plants of desired roots are grafted and damage is repaired.

10. Elite plants can be produced for kitchen garden by more than one cultivar on the same rootstock. Sometimes pollinizer is required for fruit set in such a situation it can be grafted on to shoot.

Disadvantages

1. Vegetative propagation becomes a hindrance in the evolution of new cultivars.
2. Although the asexually produced plants are precocious in bearing, but these have short life period than the sexually produced ones.
3. The size is small in vegetatively produced plants hence the dead wood can fetch little price than the huge seedling trees.

Plant Raising through Seeds

This is very simple method to produce plants. Some fruit plants like guava, phalsa and papaya are propagated though seeds. Rootstock required for vegetative propagation of other commercial fruits is also raised through seeds. The benefit of polyembryony is taken by sowing citrus rootstock seeds. For raising seedling thorough knowledge of seed maturity, seed collection and seed viability is very necessary. Cleopatra seeds are extracted along with Jattikhatti in August, than there is no seed germination in Cleopatra, but 100 per cent seeds of Jattikhatti germinate. Cleopatra seeds mature in December. Similarly Kainth (*Pyrus pashia*) seeds were collected in December and kept for stratification and sown in February. It is has been found that seeds do not require any stratification, rather seed germination is 100 per cent, if seeds are extracted in October and are sown immediately after extraction.

Raising of Nucellar Seedlings in Citrus

Most of the citrus species are polyembryonic in nature. *Citrus maxima* and Tahiti lime are monoembryonic. The number of nucellar embryos varies from species to species. The rootstock species recommended for various cultivars of commercial importance are highly polyembryonic.

To improve nucellar embryo germination, the seeds should be sown in lines at a distance of 1cm seed to seed and 5cm line to line. The seeds may be covered with fine sand and kept

Number of Seeds Per Fruit and Polyembryony in Citrus Rootstocks

	Jatti Khatti Kharna	*Khatta*	*Cleopatra*	*Pectini-fera*	*Rangpur-lime*
Number of seeds per fruit	30.56	32.54	10.32	4.36	17.62
No. of seeds/gram	15.00	7.60	10.60	10.00	18.00
Number of embryos per seed	4.60	2.53	5.66	4.56	2.14
Number of Seedlings Coming out in soil per seed	1.74	1.23	1.58	1.36	1.07

moist regularly with a sprinkler. It has been found that the practice, which is prevalent with the nurserymen, only zygotic seedlings are being used. The largersized seedlings, which are discarded, are hundred percent zygotic and smaller ones mostly of nucellar origin. Since, nucellar embryos germination is poor in the soil medium, hence remain small in size due to competition for space. For getting only nucellar seedlings, seed at full maturity may be excised and cultured in tubes having *Murashige and Skoog* medium and after germnation may be transfered in soil medium. The embryos can also be get germinated on a blotting paper as water culture.

For commercial nurseries the seeds may be sown apart and vigorous seedlings coming out first are 100 percent zygotic, which should be discarded. In practical, it has been seen that it is the zygotic embryo, which do germinate, and only the mature nucellar embryos germinate but immature do not germinate. It is particularly so when the seeds are dried after extraction. It is advisable to sow seeds just after their extraction from the fruits, for getting more number of nucellar seedlings.

Raising of Nursery in Polythene Bags

Extract mature seeds from the healthy fruits of desired rootstock species. Jatti Khatti and Kharna khatta. Fruit should be harvested in August, whereas those of Cleopatra in January. The seeds may be sown singly in the bags having prepared material for the purpose. Seeds germinate within 7-10 days. There can be 2-4 seedlings coming out from a seed. The seedling emerging first should be discarded with a sharp blade. The following seedling from the same seed shall be nucellar seedlings. Plant these seedlings by separating them individually in bags or nursery bed for final propagation.

Seed Germination

The conditions required for seed germination have appropriate moisture in the germination medium and suitable temperature for embryos to emerge. If the embryo germinates without any internal barrier, the seed is said to be *quiescent*. When a seed does not germinate even after having all appropriate conditions for the germination of seed, the seeds may be in *dormancy*. The seeds are called dormant seeds. Dormacy may be natural or chemical. The dormacy in seeds of ber and guava is due to hard seed coat, which acts as a barrier to the absorption of water and oxygen for seed germination. The presence of abscisic acid (ABA) is responsible for seed dormacy particularly in temperate fruit. In addition to ABA, some other substance like gum or tannins may be responsible for dormancy in temperate fruit seeds.

Scarification

This method is used to soften the hard seed coat to improve its permeability to water and gases. It can be done chemically or mechanically. Seed is given a beating without injuring the embryo, *e.g.*, ber, peach, etc. Seed can be soaked in water or kept in a moist place for days till it germinates. Seed can be treated with dilute acid solution for 3-4 hours to few minutes depending upon hard seed coat.

Stratification

The dormant seeds are store at low temperature after ripening treatment to keep the embryo dormant. The seeds are kept in moist sand for a number of days at 0-5°C. The seed dormancy due to ABA or other inhibitors can be overcome by placing the seeds in running water to leach the inhibitors for few days.

The grape and peach seeds showed improvement in seed germination when treated with 5000ppm of thiourea. The hormones like 100-500ppm of GA_3 for 24 to 4 hours or Ethrel (500ppm) treated seeds of peach and guava respectively showed good germination.

It has been an interesting observation that pear rootstock (*P. pashia, P. serotina, P. pyrifolia*) seed gave 100 percent germination when were sown fresh immediately after extraction during September-October (22-25°C). The earlier practice of sowing seed after stratification has been stopped. Peach and ber seeds gave good germination when were kept in moist soil during October-November and May-June respectively. The emerging seedling can be transplanted in the nursery beds at 2-4 leaf stages.

Seed Storage

Some time it becomes necessary to store the seed. Some seeds like ber, grapes, guava, citrus rootstocks; papaya, peach and pomegranate can be stored after air-drying. These seeds can tolerate some moisture loss and germinate after their storage such seeds are called orthodox seeds. On the other hand, seeds which cannot tolerate moisture loss, can only be stored for short period *e.g.* litchi and mango. Storage can be improved in controlled RH and temperature.

Cryopreservation

The embryos of seeds, *e.g.,* Coconut can be treated with chemicals like glycerol and dimethyl sulphoxide (DMSO) and preserved under *in vitro* system. The chemicals used are known as cryopectants and the method of storage in liquid nitrogen at -196°C temperature is called cryopreservation.

Embryo Culture and Establishment of Seedlings

The embryos (freshly extracted seeds) of citrus rootstocks can be excised with the help of a needle. The embryos are placed in Murashige and Skoog (1962) medium in test tubes. (Physiol Plantarum, 15 : 473 97). 5ml of E.D.T.A. solution is added to a litre of medium. The embryos excised from a seed may not be in mature state. Some embryos may remain undeveloped and do not germinate in the nursery conditions. These embryos germinate in M.S medium and seedlings can be shifted to the soil : F.Y.M and Sand medium in polythene bags.

Composition of the Culture Medium

Macroelements	*(mg)*
KNO_3	1900
NH_4NO_3	1650
$CaCl_2.2H_2O$	440
$MgSO_4.7H_2O$	370
KH_2PO_4	170
Microelements	
$MnSO_4.4H_2O$	22.3
$ZnSO_4.7H_2O$	8.6
H_3BO_3	6.2
KI	0.83
$Na_2MoO_4.2H_2O$	0.35
$CuSO_4.5H_2O$	0.025
$CoCl_2.6H_2O$	0.025
Organic Compound	
Sucrose	35g
Agar	12g
Myo-inositol	100mg
Glysine	2mg
Indole acetic acid (IAA)	2mg
Thiamine-HCl	1mg
Nicotinic acid	0.5mg
Pyridoxine	0.5mg
Kinetin	0.4mg
Volume raised to one liter with the addition of distilled water.	

Iron (Fe) as E.D.T.A

$FeSO_4.7H_2O$	5.57g
Na_2 E.D.T.A	7.45g
Distilled water	1 liter

Micropropagation

Just like embryo culturing, here a tissue can be used for growing under controlled nutrition and environment. The explants varies from fruit to fruit. Explants may be tip culture, and ovule culture. Tissue culture technique has been perfected in banana, Strawberry and peaches. It is a very costly method since equipment required is costly. It is also very technical. The plants are to be hardened before these are planted in the filed.

Raising of Rootstock for Pear

Pear is commonly propagated on pear root suckers (*Pyrus calleryana*) or called 'Batank', or on seedlings of kainth (*Pyrus pashia*). In earlier publications pear root suckers have been considered to be Pyrus pyrifolia, which has been proved wrong. No root suckers emerge from own rooted patharnakh (*P. pyrifolia*) or shiara (*P. serotina*) cutting when used as rootstock. The Kanith roots also do not produce any root suckers; it produced collar suckers only which cannot be used for propagation. A twenty-year-old tree raised from root sucker did not come into flowering Shiara bearing, under condition whereas Kainth Shiara and Patharnakh tree bear profusely. The root suckers are usually removed from the orchards during October-November. Plant the suckers in beds at 15cm apart in lines at 30cm distance. Only surviving suckers should be grafted during January.

Cuttings

For raising rootstock from patharnath (*Pyrus pyrifolia*); Shiara (*P. serotina*) or kainth (*P. pashia*) cuttings, the cuttings are prepared during December. To improve the rooting percentage the basal portions of the cutting may be dipped in 100ppm IBA for 24 hours, before planting.

Some other fruit plants can also be propagated through cuttings *e.g.* Pomegranate, grapes, guava and plum. The cuttings are prepared from one-year-old shoots in December and treated with IBA solution for improving rooting. The white ant attack is checked by using chloropriphos (20%) 10ml/litre 10-15 days after planting and again 20-30 days after. The cutting beds are kept moist for maximum rooting.

Layering

This is an easy method of raising well-sized plants. Roots are induced on the shoots on the mother tree itself. There are many methods of layering like air layering, mound layering, trench layering and serpentine layering, etc. Only air layering and mound layering are commercially used.

Air Layering

One-year-old 40-50cm long straight shoots are selected. Ring of bark 3 cm wide just near the bud is removed. Moss grass or crushed Echorinia roots are used to wrap the naked part and few centimeters further, with the help of polythene sheet. July-August is the best time for air layering in litchi or guava. Use of IBA 100ppm at the site of the ring removed can produce early and good number of roots in a layer. The layers should be removed from the parent tree, as and when roots are visible in the polythene sheet or after 40-45 days of air layering. The rooted air layers are planted in the nursery after removing the polythene sheet. After a year or so these plants are planted in the field.

Mound Layering

(Stooling) this method is very successful for guava propagation. The mother plants are established at one meter apart. The main shoot is cut during March at a height of 10cm to initiate 7-10 fast growing shoots. From these shoots bark as in air layers is removed at 5 cm above the ground level during July-August. Then earthing up erects a mound, to cover at least 15-20, cm of the shoots length and the soil is kept moist. Roots develop with in 20-25 days above the ringed portions. The rooted shoots are detached from the mother plant below the roots during January-February, after gentlly removing the soil. The rooted shoots are planted in the nursery. The same mother plants are used for a number of years for raising the mound layers.

Topophysis

Juvenile and mature phases coexist on an adult tree. If bud wood/graft wood is taken from the basal part (near to the root) of the mother tree, the plants produced shall remain vegetative (juvenile) for longer period than those propagated from the upper portion of the same tree. Therefore to achieve precocity in bearing bud-stick should be taken from the upper parts of the mother plants. This difference in juvenility is termed as topophysis.

Budding

In budding only one healthy bud is taken as scion from the scion wood and inserted in the bark of a pencil thick rootstock seedling (shoot). Budding is usually carried out in spring and rainy season. It is the best method to propagate citrus, guava, ber and peaches. The mother plants of desired cultivars are selected and healthy bud wood is taken at the time of budding. The bud sticks are kept in polythene sheet by wrapping in moist moss or gunny bag piece. Only healthy rootstock seedling should be budded at a height of 15 to 20 cms above ground level. The budding type may be T-budding, patch budding or ring budding depending upon the success of the method for a particular species. In budding, bud has phloem and little xylem attached to it. After the insertion of the bud onto the exposed xylem, callus develops from rootstock xylem and holds the bud. Then cambium union take place and bud sprouts to produce a plant.

Grafting

In grafting more than one bud on a small shoot 5-10 cm long from the mother plant is taken as scion and joined properly on the rootstock. Grafting is common in mango and temperate fruits. Grafting is of many types *vis tongue*, cleft, side, veneer and whip or splice grafting. In mangoes side or veneer grafting is usually practised, whereas in temperate fruit plants tongue and cleft grafting is carried out.

Freshly cut graft wood when brought into intimate contact of the rootstock by any of the above-mentioned methods, brings activity in exposed phloem and xylem cells. Cells in the cambium of the scion and rootstock first intermingle and interlock and then differentiate into new cambial cells, which produce a new vascular tissue. Buds on the scion sprout to produce shoots and a new plant is formed.

Methods of Grafting

Approach Grafting

The scion shoot is tied with the stock by taking the rootstock in a polythene bag or pot to the scion tree. After the union takes place, the scion is removed from the mother scion tree gradually. Similarly the rootstock shoot above the graft union is severed. Various types of approach grafting are :

Inarching

The rootstock is taken to the mother scion tree. A thin slice of bark is removed from the rootstock shoot at a height of 20 cm or so. A similar cut is given on the scion. The cambium layers of both stock and scion are brought together in a matching manner and tied firmly with the help of polythene strip or 'Seba'. It is the common method employed for the propagation of mango. Inarching is useful in rejuvenation of old or inferior mango trees.

Tongue-Grafting

The rootstock is cut as in-inarching and a downward cut is given to form a tongue facing upward. Similar tongue is prepared on the scion in such a manner that both the tongue fit in and cambium get in touch firmly, and tied as in inarching. This is most commonly practised in temperate fruits pear and plums. Tongue-grafting can be carried out as inarching and by taking scion from the mother tree and grafting it on the rootstock seedlings in the nursery.

Whip/Splice Grafting

This can only be carried out when thickness of the rootstock and scion is almost the same. One-year-old seedlings are given a diagonal cut from down to upward at a height of 20 cm or so. The length of the splice should be 3-4 cm. The top of the rootstock is cut. A similar splice is removed from the proximal end of the scion. Both cut surface are brought together to match each other and both are tied tightly with a polythene strip.

Side Grafting

The rootstock is prepared by removing a bark and the scion is prepared as in splice grafting from both sides and inserted under the bark and tied.

Veneer Grafting

The only difference in veneer and side grafting is that in veneer grafting the scion is prepared by giving a slanting cut 3-4 cm long only from one side where as in side grafting it is prepared from both sides. Veneer grafting is most successful method employed for grafting mango during August-September. In veneer grafting, the rootstock seedling is given 4-5 cm long cut on the bark as well as small portion of wood is removed to expose the wood which should match with the thickness of scion. The scion is placed and stock bark is replaced over the scion. Thus the scion do not slip further. Both cambium are made to match each other. The major advantage is scion can be transported from a distant place.

Cleft or Wedge Grafting

If the rootstock thickness is more than the scion then cleft grafting is carried out. This is mostly used in pear propagation. The stock is decapitated at 20-25 cm height and a cleft is made in the centre of stock. The length of this split should be 4-5 cm. A scion is prepared to fit in this cleft by giving slanting cuts on the both sides to form a wedge. The scion length may vary from 7-10 cm. The scion is made to fit in the rootstock so as one side of the scion should match the stock from out side. Sometimes two scions can be inserted in a thicker rootstock.

Stone Grafting

The germinating stones of mango are splice or wedge grafted with a mature scion. A slanting cut is made on the epicotyl part of the seedling with a matching cut on the proximal end of the scion and are tied together. Under North Indian conditions, this method could not be popularised.

Saddle Grafting

The rootstock is beheaded at a height of 20-25 cm and upward sloping cuts are made opposite each other forming a sharp end of the stock. On the scion at proximal end a tongue is made so that the wedge of the rootstock may fit in the scion tongue. Both pieces are tied with polythene.

Double Working

Sometimes it become necessary to introduce a piece as interstock other than the scion and stock species. The interstock should be compatible with both the scion and stock. This can be exploited to produce dwarf and precocious pear plants. Patharnakh as interstock has proved well between *Pyrus calleryana* rootsuckers and soft pears like Le'conte or 'Baggugosha'. An other example is Old Home or Hardy pear as interstock in *Bartlett* pear and quince (*Cydonia oblonga*).

Budding Methods

Vegetative buds are used as a scion. The time of budding depends upon the availability of buds and cell sap flow in the rootstock seedlings so that bark separates easily to take scion bud. The different methods of budding :

T-Budding

The scion is removed from the shoot in the shape of the

bud. The bud may or may not have small portion of xylem wood. A T-shaped cut is given in the rootstock just as deep as the bark. Then the bud is inserted under the bark in this T-shaped exposed xylem and tied with a polythene strip. The portion of the active bud is left uncovered for sprouting. This is the most important method of budding in citrus, peaches and roses.

Inverted-T-Budding

The only difference is that inverted T-cut is given on the rootstock seedling and scion bud is inserted from below upward.

Patch Budding

A rectangular patch of bark is removed completely from rootstock seedling of sufficient thickness at a height of 20 cm. A similar-sized patch bud is taken from the scion and placed over to the stock and tied. It is a little time consuming to perform. However patch budding is very successful to bud guava and aonla.

Ring Budding

The rootstock is headed back just as in grafting at a height of 15-20 cm. A ring of bark is removed from the top of the beheaded rootstock seedling. A similar ring along with a bud is taken from the scion and inserted on to the rootstock. This is commonly done in peach and mulberry.

Chip-budding

It is carried out when cell sap flow is too low and budding is not possible with other methods.

Forkert Budding

It is just like patch budding, but patch is not totally removed from the stock. It is left attached from one side. While placing a bud on the patch, the bark patch is again placed on the bud by removing a little portion in flop to sprouting of bud.

7

ROOTSTOCK PROBLEMS AND THEIR MANAGEMENT

In early times, the fruit plants were raised from seeds. These plants had a long juvenile period and there was little or no uniformity in tree growth, yield and fruit quality due to their zygotic origin. In modern times, to reduce the juvenile period and bring uniformity in plant vigour, yield and fruit quality, the plants are being propagated through vegetative means. Thus, either the plants are raised through rooting of cuttings, *e.g.* grapes and plums, or by the use of a rootstock for a particular scion cultivar. The rootstock once used for the propagation of a fruit plant can not be changed later on therefore, the nursery man should have ample knowledge pertaining to the behaviour of different rootsock for a scion in a set of soil and climatic conditions.

To make recommendation for the use of a rootstock for a particular fruit plant, it requires long-term experimentation. It is difficult to make a universal recommendation of a particular rootstock for a particular cultivar. For example, for citrus kharna Khatta is the most prevalent used rootstock. In Punjab, it is a failure due to exocortis infection. In Punjab, it is the Jatti Khatti which has been considered the best rootstock. Here too Jatti Khatti has sent the Blooded cultivar of sweet oranges to decline, due to crease formation at the bud-union. Hence, different rootstocks for different cultivars are being evaluated and have been recommended. Now Cleopatra has been recommended for Blood Red sweet orange and Pectinifera for Mosambi. For high density plantations of kinnow *Poncirus trifoliata* can be successfully exploited. This proves beyond doubt that time has come to give rootstock recommendation for individual cultivars of sweet orange and mandarins. Troyer and Carrizo citranges which have proved good in U.S.A. have failed to perform in India due to exocortis susceptibility.

Factors Affecting Rootstock Recommendations

1. Soil Factors

Soil is a natural gift to human beings. Due to raised situation/compulsions, all types of soils are put under fruit cultivation. The factors like soil structure, organic matter content, concentration of salts and soil pH greatly influence the recommendation of a rootstock. For example, Kharna Khatta is being recommended for kinnow in heavy soils of Uttar Pradesh, and Rangpurlime for Nagpursoils. Similarly *Pyrus pashia (kainth) and P. serotina* (Shiara) seedlings perform better in mountaineous and rocky soils for pear. On the other hand, Pear root suckers (*Pyrus calleryana*) oriental pear is a better rootstock for plains and waterlogged soils. Therefore one cannot make a blanket recommendation of a particular rootstock for all soils. Similarly M_9 and M_{27} dwarfing rootstocks for apple have failed in India due to poor soil fertility of the area.

2. Climate

The rainfall influences the climate the most. Rainfall varies from state to state and even within the state. Some citrus rootstocks perform better in high rain fall conditions than arid or arid irrigated region. Jatti Khatti do not develop phytophthora symptoms in arid or semi-arid climate of Punjab and Rajasthan, while in Amritsar and other central districts of Punjab and moist conditions of Haryana and Uttar Pradesh it is easily infected by phytophthora fungi. *Poncirus trifoliata* performed well in Hoshiarpur belt than in Abohar due to temperature variations.

Own rooted cuttings of Patharnakh (*Pyrus pyrifolia*) trees were prone to tree felling in areas with heavy rain fall than the patharnakh or softpear plants raised on pear root suckers (*Pyrus calleryana*).

3. Cultivar Differences

The rootstocks recommended for sweet orange cultivars cannot be recommended for mandarins. Even the cultivars perform differentially in a species on a particular rootstock. Jatti Khatti (*Citrus jambhiri*) showed a crease formation at the bud union with Blood Red cultivar for which a new rootstock cleopatra has been recommended. Similarly, Pectinifera is being recommended for Mosambi. Pineapple and valencialate cultivars are doing well on Jatti khatti. In pear also Bartlett cultivar is not compatible with quince (*Cydonia oblonga*). For Sathgudi sweet-orange Sathgudi is being used as a rootstock, but it is not doing well with Mosambi and Pineapple oranges.

4. Regional Adaptations

Some rootstocks perform good in one region and if used for other regions the performance is not upto mark. Citranges particularly Troyer and carrizo are performing very well in U.S.A. but in India particularly Punjab these rootstocks have performed very low. Similarly, Rangpur lime is recommended for citrus cultivars in south is doing very well but in Punjab it has failed totally.

5. Pests and Disease Incidence

Some of the rootstocks can tolerate the attack of certain bioagents but some cannot tolerate the attack to a great extent. Tristeza susceptible sour orange is not being used as a rootstock for citrus. Similarly Dogridge and Salt creek rootstocks of grapes are resistant to nematodes attack than the other rootstocks.

Benefits

1. Nutritional Benefits

Certain rootstocks help the scion cultivar to uptake and build higher quantities of nutrients in the foliage than others. For example, Patharnakh plants raised through rooting of cuttings (without any rootstock) show acute die back symptoms of Zinc deficiency, whereas patharnakh grafted on root suckers

(*Pyrus calleryana*) oriental pear had no such problem. In an other experiment four cultivars, *viz.* Smith, Leconte, Kieffer and Bagugosha were grafted on two different rootstocks, *viz.* Pyrus pashia (Kainth Seedlings) and oriental pear rootsucker (Pyrus calleryana). It was observed that N and P. amounts in the leaves of scion cultivars were more on pear rootsuckers than on Kanith, rootstock. K was more on Kainth than pear root suckers.

The different stionic combinations in mandrins resulted into better NPK contents in the foliage of scion cultivars on Jatti Khatti (*Citrus jambhiri*). But the combinations with sweet orange cultivar showed that Troyer and carrizo rootstocks had higher NPK contents than Jattikhatti.

2. Plant Vigour

The rootstocks influence the growth of fruit trees much more than any other character. In apple, pear and citrus the examples are in plenty. Pear trees on D_4 and Kainth rootstock were the most vigorous and on quince (*Cydonia oblonga*) the dwarfest. Similarly citrumello and Kharna Khatta had bigger trees of sweet oranges than Rangpur lime, Cleopatra and Trifoliate orange. In ber the *Z. numularia* had small ber plants than *Z. mauritiana*. In apple M_9 is the most dwarfing, M_7 semidwarf and M_{11} and M_{16} most vigorous.

3. Influence on Anchorage

The scion depends upon the rootsystem of a rootstock for anchorage, sometimes many cultivars have very good fruit characters, but have poor root system, for example, mango cultivar Kishan Bhog has very good fruit but seedling growth is poor hence such cultivars need a rootstock for grafting. Pear on quince rootstock need support to keep erect. Similarly own rooted Patharnakh cuttings trees are prone to tree felling during rainy season because of poor anchorage. In citrus Kharna Khatta and Jatti Khatti have vigorous root system than Citranges Troyer and carrizo and trifoliate orange.

4. Juvenility

Dwarfing rootstocks usually bring early bearing in fruits

by reducing the juvenile period. Pear is precocious on quince C, apple on M_7 and M_9 Citrus on Citranges and ber on *Z. nummularia.*

5. Influence on Flowering and Fruiting

This is the most pronounced effect of a rootstock. The flowering is advanced by few days on dwarfing rootstock than a vigorous one. Usually yield per tree is more on a vigorous tree than on a dwarfing rootstock. Pear on quince produced only 5 kg. of fruit in comparison to 250 kg. on D_4 and Kainth rootstocks. Kinnow can bear over two thousand fruits on Jatti Khatti in comparison to 300-400 on trifoliate orange.

6. Fruit Size and Quality

The fruit of Kinnow are larger in size on Kharna Khatta than on cleopatra and trifoliate orange. The fruit quality is better on trifoliate orange than Kharna Khatta. The bitterness in grape fruits is absent when budded on Trifoliate orange rootstock, but it persists in plants propagated on Jatti khatti (*Citrus jambhiri*) and other lemons as rootstock. Coorg mandarin fruits are of better quality on cleopatra than on Rangpur lime and Troyer citrange.

7. Influence on Fruit Maturity

Almost all rootstocks which have dwarfing effect on the trees, bring fruit maturity and fruit ripening a few days earlier than the trees on its seedling rootstock or on vigorous rootstocks. Hence, this character can be exploited for extending the harvesting period. Kinnow ripens earlier on Trifoliate orange, followed by citranges, cleopatra and Jatti khatti. Thus kinnow marketing can be carried out from November to end February.

8. Resistance to Some Diseases

In citrus cleopatra root stock is resistant to many diseases particularly is phytophthera and is salt tolerant. Rough lemon (Jatti khatti) is tolerant to tristeza. Nemaguard Flordaguard rootstocks for peaches is resistant to nematode and crown gall, in mango kurukkan rootstock is salt tolerant. In grapes, Dogridge is resistant to Phylloxera and nematodes. Lucknow-49 a guava rootstock have shown resistance to guava wilt.

9. Toleration to Winter Injury

Rootstock can also impart resistance to the scion cultivar against frost and freeze injuries. Sucking mango as rootstock has better toleration to cold injury in Dusehri mango than Dusehari seedling root stock. Cleopatra is also regarded as freeze tolerant for citrus. Trifoliate orange is very hardy rootstock for winter.

What is Desired

In citrus the rootstocks have to be recommended region wise. The nucellar embryos should be cultured in growth media and uniform material should only be used in each rootstock. New rootstocks may be evaluated for dwarfing and tolerant to viral diseases.

In mango seedlings of mango variety which is tolerant to frost/freeze should be evolved/evaluated. A dwarfing rootstock is the need of hour for Langra, Dusehari and S.B. Chausa cultivars.

For Guava a rootstock resistant to wilt is required. The rootstock should be tolerant to fruit fly. In ber it is very important to develop a rootstock which should be dwarfing and can impart resistance to powdery mildew.

In pear a semidwarf rootstock with precious in nature is required. Peach and plum need good rootstock to have anchorage and dwarfing effect on cultivars.

8

ORCHARD PLANNING

Orcharding is a long-term investment hence, require planning. Any mistake committed in laying out, planting distance, fruit plant/cultivar choice, procurement of plants, own nursery raising, irrigation system/method shall not be undone without much loss. The efficiency of the orchard shall be based on insite of the Orchard planner. While planning help of the Orchardist all ready in the profession and horticulture expert can not undermined.

The following components should be given due attention.

Fencing and Wind Break

A strong fence with the use of barbed wire should be provided around the field meant for planting of Orchard. Months to a year before actual planting is done karonda (*Carissa carandus*) or Jatti khatti (*Citrus Jambhiri*) can be planted near the barbed wire. This will provide impenetrable and strong fence, while protect the Orchard against the entery of wild/stray animals.

Plant Eucalyptus or Jamun seedlings at a proper distance inside the provided hedge to provide the Orchard a strong wind break. A trench one meter deep may be dug 2 meter away from the wind break to check the wind break plant roots interference with the fruit plant roots in the latter years.

Land Preparation

Before actual planting of the fruit plants the land should be cleared of all bushes and trees. Remove the complete root systems of the uprooted plants to check their resprouting later on. After levelling and making blocks green manuring by sowing dhiancha, sunhemp, guara or cowpeas should be done. This practice shall provide the land adequate organic matter and aeration, required for fruit plants.

In arid and foot hills the lands are usually undulated. In such situations, the land may be divided into blocks before levelling. Level the individual block only, this shall avoid unneccesary lifting of upper soil from one point to the other.

Roads and Paths

A well laid out network of paths and roads shall be useful throughout the life of the orchard the paths should be provided with the plantation of Khabal grass which will keep the dust intact and help in the penetration of rain water fast. Such paths shall help in controlling the erosion of soil by wind and water.

Tubewell/Tank

The digging of well or a tank shall be useful in areas where irrigation water is available only when canals are providing it. Water for use can be lifted from the tank/well when ever needed most. Area for boring a tubwell or a tank may be marked at an appropriate place in the orchard land.

Buildings

In a big-sized orchard some buildings are required, to store implements, place fruits after harvesting, for resting of labour and a rest house for the owner. These buildings should be prepared before hand.

Layout

Layout of an orchard is carried out from road or path side adjoining the area to be planted under fruits. Layout is done to mark the positions of the plants in the field for digging and filling of pits. It is a technical job hence the help of a horticulture expert should be requisitioned before hand any mistake made at this cannot be corrected later on. Orcharding being a long term venture, therefore distance for planting for a particular fruit should be clear to the grower. The drawing of planting plan on a particular system may be made on paper by making provisions of irrigation channel and leaving paths for easy movement of men and vehicle at a later stage of orchard development.

Planting Systems

There are many systems being followed for planting of an orchard depending upon the soil and topography.

1. Square System

This is most popular system of planting being followed by the orchardist through out the country. The distance between plant to plant and row to row is the same and plants are at right angle, thus four plants forming a square are seen through out the field. The movement of machinery is uninterrupted in this system.

2. Rectangular System

This is similar to square system with only difference that line to line distance is more than the plant to plant thus forms a rectangle.

3. Triangular System

In this system trees are planted as in rectangular but the plants in 2nd, 4th and other alternate rows come midway of the plants planted in 1st, 3rd and so on rows. Thus there will be one more plant in odd rows than even number rows. Interculture becomes difficult in this system, which can be carried out in diagonal positions.

4. Hexagonal System

In this system plants are planted by forming equilateral triangles. The distance between the plants remains the same, but row to row distance is reduced. This accommodates 15 percent of more plants than the square system. Generally trees having less height are planted on this system example, citrus and plums. Litchi, mango and pears should not be planted by this system, since at maturity of the trees, branches get intermingled.

5. Quin Cunx or Diagonal System

This is first a square system, by planting one additional plant in the centre of the square. The filler trees having less life than the main fruit trees is planted as additional plant, for example kinnow or plum is planted as filler in mango and litchi orchards. The fillers are removed when main plantation occupy the whole space and fillers start intermingling with the main plants.

6. Contour System

The orchards are planted by this system in hills or foot hills. To reduce the chance of soil erosion contours with almost level land are formed. The layout for contours is started from the lowest contour forming a base line and carried upward. In the foot hills more or less levelled areas are separately marked by forming strong ridges to hold water.

7. Hedgerow System

In this system plants are planted very closely in rows at a proper distance. This system is being followed for guava and pomegranates. Row to row openness provide the desired aeration and movement in the orchard. The yield can be increased per unit area on this system by adopting some techniques.

After selecting the layout system, the base line is marked by establishing long poles along the path, road or any other permanent structure. From this line by leaving one half of the planting distance from the base line another line is established. The position of the first plant is marked on this line and subsequent points are marked by actually measuring the distance from plant to plant with a measuring tape. Another base line is marked along the other edge of the field at right angle to the first base line or the first marked plant. Position of the plants on the second line and subsequent lines are marked from this base line. The long poles can be established at desired distances on the second base line. Four men are required to complete the layout. One keeps the rope/tape, second hold the pegs, third pushes the peg at proper site and fourth to align the rows. The right angle at the first base line can be drawn with the help of a measuring tape. One end of the tape is kept at a point on the first base line and tape is stretched along line upto 3 metres, then 4 metres on the second base line and adjusting 5 m to the point of start. This will form a right angle triangle. The second base line is thus adjusted accordingly. Thus positions of all the plants are marked with the help of pegs.

Digging and Filling of Pits

A planting board of 3 feet length having V shaped cuts at both ends and at the centre is used for marking the area for digging pits. Two guide pegs are inserted at the ends of the board and the marked peg is removed. Approximately one metre deep pits are dug. If there is hard pan below, more deep pits can be dug to remove the hard pan. Upper 50 cm top soil is kept on one side and rest on the other. Top soil is mixed with at least a quintal of farm yard manure or 50 kg of vermicompost and pits are filled with this mixture. After filling the pits irrigation is applied. The uneven pits are levelled by using the discarded soil.

Planting

The desired soil is removed from the centre of the pit. The earth ball of the plant is placed in this hole in such a manner that upper surface of earth ball is at same level with the soil. Place the plant in the centre by using the planting board. The plant trunk should adjust in the V at the centre of the board. Loose earth on all around the earth ball is pressed by using both the feet by standing carefully along the plant. Keep the feet away from the earth ball so that it should not get damaged. Apply irrigation and relevel the surroundings.

Planting Distance

Planting distance depends upon many factors such as soil fertility, cultivar, rootstock used, and training system to be followed. To maintain high productivity of the orchard, it is essential to provide, required planting distance. The plants with less space grow poorly and get intermingled among each other at an early age and are attacked by pests and diseases, due to poor light penetration. While planting an orchard planting distance should be carefully discussed with the expert in the field. Varied planting distances are followed by the growers in the field for a particular fruit depending upon the area. A typical example is Kinnow, it is being planted at 10′ × 10′; 10′ × 20′; 15′ × 15′; 17′ × 17′ and 20′ × 20′. However, Punjab Agricultural university has made recommendation on planting distances for various fruits which need to be followed.

Fruit Tree	*Planting Distance (m)*	*Number of plants/hect. on square system*
1. Mango		
(*i*) Langra	11 × 11	81
(*ii*) Dusehri Alphonso, etc.	10 × 10	100
(*iii*) Amrapali	9 × 9	121
2. Litchi	7 × 7	196
3. Jamun	10 × 10	100
4. Sapota	10 × 10	100
5. Pear		
(*i*) Patharmakh	7.5 × 7.5	169
(*ii*) Soft pear	7 × 7	196
6. Guava, Peach, Loquat	7 × 7	196
7. Citrus and plum	6 × 6	272
8. Pomegranate	5 × 5	400
9. Grapes		
(*i*) Head system	2 × 1.5	3350
(*ii*) Kniffin system	3 × 3	1089
(*iii*) Bower system	3 × 6	561
10. Phalsa and papaya	1.5 × 1.5	4296

Planting Time

Both evergreen and deciduous type of fruit trees differ in their planting time.

Evergreen

Planting of evergreen fruit plants can be done in two seasons, *viz.* spring (February-March) and rainy season (July to mid November). Mango, litchi, sapota and citrus (kinnow) is usually planted in the rainy season. Experience in planting shows that mango and litchi plants perform better when planted from September to mid-November than early planting in July August. Ber, guava, Jamun, etc. can be planted both in February-March and rainy season planting of mango and litchi during spring season should be avoided. Well-sized guava plants can also be planted bear rooted (without earth balls) in September-October, Apply irrigation immediately after planting of evergreens. If light and frequent irrigation is applied to evergreens planted during July-August, it helps in easy establishment of the plants. In kinnow one can get at least two flushes of new growth upto December.

Deciduous Plants

These plants are usually planted in dormant stage with bear roots during winter. The most suitable time for planting these plants is from mid-December to mid-January. However, grapes, phalsa and soft pears can be planted from mid-January to mid-February. The summer dormant ber plants can also be planted bear rooted in December-January.

Procurement of Plants

The growers should visit the nurseries for making a comparison of the health of the plants, before getting the plants booked with a particular nursery. It is not necessary to purchase the plants only from Government or university nurseries. The focus should be on health of the plant and purity of scion cultivar. It is very important to verify from the manager of the nursery the rootstock used for the propagation of plants. Book plants with the selected nursery at least a month in advance of their picking/taking delivery. Book at least 10 percent more than the actual requirement and purchase. These plants should be kept at the planting site for filling of gaps which are likely to occur. The evergreens should be brought in plastic bags of 20 × 30 cm size, whereas deciduous plants should be kept in the nursery to fill the occurring gaps. These plants should be given due care as that of field planted plants.

A certificate from the nursery manager should be taken for purity of scion and rootstock. In case of citrus it should also be got certified that the plants are free from virus. Moreover, the age of the nursery plants should not be more than one or one and half year. While digging the plants along with earth, it should be assured that sufficient number of feeder and anchoring roots are intact. Similarly while purchasing care should be taken that over 80 percent of feeder and tap root should be dug from the nursery. It has been seen in pears propagated on Kainth seedlings (*Pyrus pashia*) that plants with 10-15 cm of tap root got killed in the field during April-May after sprouting. Hence, the deciduous fruits should have a taproot upto 20-25 cm long with feeder roots.

Transportation

One should transport the plants in his own hired vehicle. Plants particularly evergreens should not be lifted in tractor trollies. The plants should be carefully loaded in the vehicle and closely packed. It will be beneficial to have the plants packed in crates or baskets.

Care of the Plants on Arrival

Handle with care the plants while unloading individual plant. Keep one hand below the earth ball one on the earth ball. Don't pick from the trunks of the plants, the weight of the earth ball can cause damage to feeder roots. However, if the plants are in packs or baskets, then these are picked from the containers itself. Place the plants in shade preferably in varandah. Sprinkle water if required. One can keep the plants for a few days in shade before being planted.

Care After Planting

While planting the plants should not be picked from the rootstock trunks. Hold the earth balls in both the hands to avoid earth ball breakage. Earth balls should be treated like eggs.

Newly planted plants being very tender and under tremendous shock need attention. Apply irrigation just after planting. If the plants need support, supply the same immediately. Inspect the plants regularly for rootstock sprouts, removal of plastic material from graft/bud position. It will be better if 10-15 cms of height of the scion is cleared of young shoots as well as leaves. After planting of the plants white ants do attack plants within a month or so. Therefore, it is very necessary to apply 500 ml of solution of chloropyriphos 20% should be added to one litre @ 10 ml/litre to each plant.

The young plantation should be protected from harsh winters as well as summers. Fill the gaps if any the moment it occurs in evergreen and after a year in deciduous plants.

9

CANOPY AND FRUIT MANAGEMENT

Canopy management is one of the most important fruit plant management practice. This forms the basis of precociousness and longevity of the fruit trees in an orchard.

The strong scaffold system help the trees to produce heavy crop loads of quality fruits without any major limb breakage, example, patharnakh. Some trees bear regularly only if pruned every year, example, peach and grapes. Regular annual pruning is essential for the induction of good amount of new growth, which will provide the tree maximum fruit brearing area uniformly over the whole tree. Unpruned trees grow beyond limits provided for each tree in an orchard. The branches start intermingling with the branches of adjacent trees giving an impression of a jungle. The photosynthetic activity of the trees get adversely affected leading to the poor yields of low quality fruits. Orchards become economically unproductive. To keep the fruit trees within the area provided for their spread for whole of the life, canopy managemnet is of prime importance.

Canopy management practices are discussed below :

1. Training of Fruit Trees

After planting the fruit plants in the orchard, the training starts from day one. Initially few branches arising from rootstock portion and 10-15 cms above the union are removed at the time of planting. When vines are staked or tied over a trelly or pergola in a certain fashion or some of the parts are removed with a view to give it a desired shape, the operation is called training. The training is done with the following objectives.

1. To admit light upto centre of the tree and provide sufficient movement of air across the plant.
2. To increase photosynthetic activity by exposing leaves to the sun.
3. To provide strong scaffold system which could bear the heavy load of fruits, without limb breakage.
4. To make hoeing, spraying, irrigation and other cultural practices at a nominal cost.
5. To get balanced distribution of fruit over the tree.

Training Systems

The following training systems are normally followed for tree crops.

(*i*) Central Leader System : The central leader is allowed to grow uninterrupted. The secondaries grow on the central axis on all directions. The fruit tree grow in a natural way. The tree trunks become very strong due to the spread of many scaffolds and secondaries. The trees become tall and spread mostly unmanageable at maturity. This system is most suited to litchi and mango.

(*ii*) Modified Leader System : The central leader is allowed to grow to produce 3-4 side branches, then it is headed back at 75 cm height for low headed and at 90 cm for high headed plants. In the next year, the top bud sprouts to take the shape of the central leader, which is again headed back after getting 2-3 scaffolds at the last scaffold giving it an open centre. This can be done after 2-3 years of removal of the central leader, that is why the system of training is called modified leader system of training. In all, there can be 5-7 scaffolds on which secondaries are made to develop by removing apical dominance of each scaffold periodically. Thus a tree takes the shape of an umbrella in spread and a cone in height. Modified leader system trees possess a strong durable framework like central leader system and openness of the open-centre system. Tress do not grow as tall as in central leader system, thus remain manageable for a long time. Orchard efficiency is never adversely affected due to overgrowth. There is no breakage of limbs due to the load of the fruits. This system is suited to pears, peaches, guava, mango, litchi and many more which can grow to form big trees.

(*iii*) Open-centre System : The plants are planted in the orchard and simultaneously headed back to 75 cm height. The well placed 4-5 side branches are allowed to develop on the main axis. The top growing axis is again cut and is not allowed to resprout and give side branches. The selected scaffolds are made to produce secondaries and tertiaries just like in modified leader system. Thus the tree gives the appearance of the umbrella. The limbs can cause breakage on the small main axis due to over weight of fruit on few scaffolds in comparison to modified leader system. This system is preferred for those fruit trees which have 10-15 years of life span example peaches and plums. This system is very suitable for exotic peaches in North Indian plains.

Pruning

Normally pruning is an invigorating process. Many a times it is carried out to encourage new growth and fruiting. Pruning is defined as the removal of unwanted parts, *viz.* shoots, branches roots to allow the fast growth in the remaining parts.

Objectives

(*i*) To remove the apical dominance for encouraging branching.

(*ii*) To remove unproductive over crowded branches.

(*iii*) To remove diseased and dead wood branches.

(*iv*) To encourage vegetative growth.

(*v*) To control the overall size of the fruit tree.

(*vi*) To regulate fruiting for regular cropping.

(*vii*) To give particular training.

Method of Pruning

Annual pruning can be done in two ways :

***(i)* Heading Back :** This type of pruning can be done in both evergreen and deciduous fruit trees to remove apical dominance and encourage side branching. For peaches which bear on new growth, this type of pruning is an annual feature for getting regular fruiting. Normally 1/3 of the top shoot is removed every year during pruning. However, in some fruits like phalsa the whole bush is headed back to the ground level to develop sufficient number of branches for bearing regularly.

***(ii)* Thinning Out :** When there is a bushy growth of side shoots on the secondaries or tertiaries, some of the branches are removed entirely from point of emergence without leaving any stub. It results in providing light and aeration in the tree. Thinning out encourages fast growth of the remaining terminals. This gives the tree a laggy growth. To get best results from pruning a mix of heading back and thinning out will be best for long-term production of quality fruits from peaches.

***(iii)* Rejuvenation :** Old, tall overgrowing fruit trees of mango litchi, pear, etc. can be rejuvenated to get the fruit from the same trees for a number of years again. In literature it has been mentioned that the trees may be rejuvenated in parts, *i.e.* head back few scaffolds in one year and few is second and rest in the third. This way a tree shall continue to provide fruit for these years and rejuvenated simultaneously. It has been found practically wrong. To rejuvenate the tree whole of the tree should be headed back at one time. Keep only 4-5 well placed scaffolds to a length of 15-20 cm. Many sprouts appear on these stubs. Make selection in the same season and keep 7-10 scaffolds on all the stubs not more than two on each stub in any case. Such trees come into bearing again in 3-4 years after rejuvenation. The rejuvenation has been practised in the field for pear, mango and guava orchards successfully. The time of rejuvenation for pear is December-January, for mango-January-February and for guava March-April and August-September.

Time of Pruning

Time of pruning in different fruit plants differ from fruit to fruit. Normally deciduous fruits trees are pruned when complete dormant after shedding of leaves. Pruning of pear, peach and plum should be done in December-January, whereas phalsa and grapes needs to be pruned end January-February first week. Ber which is summer deciduous should be pruned in May-June.

2. Bending of Scaffolds

To manage the canopy and get early fruiting from trees with long juvenile period, the bending of branches have been successfully practised in pear. The bending of flexible scaffolds downward can be carried out by tieing the scaffolds to the trunk of the same tree or on to pegs in the basin of the tree. Bending of scaffolds provides the advantage of geotropism. The spur formation is enhanced by two years. Normally it becomes difficult in getting secondaries on the main scaffolds in pear training. Bending also helps in the sprouting of buds to produce good number of secondaries on a scaffold. Thus training on modified leader system is enhanced by one to two years. The bending of scaffolds can be carried out in Guava, Mango and litchi in addition to pear.

3. Provision of Supports

The selection and breeding programme have developed precocious and heavy bearing cultivars in different fruits. Well managed orchards bear very heavily in their early years of bearing. The tender scaffold system is unable to bear this heavy weight and limbs tend to break from the centre. Some limbs break out rightly from the point of union on the trunk. This is particularly so in pathanakh pear, Peach cultivars and kinnow. The patharnakh orchards get devastated due to limb breakage these orchards never come to that position in the coming years. Only the few cms long broken limbs continue to bear fruit. Peach tree break from the centre, the broken trees usually die. To save the hard developed canopy of tree, the bamboo supports should be provided to the major heavily bearing limbs/shoots. This is required to 3-4 years of early bearing. Later on tree establish itself with required yield/fruit weight.

Training of Grapes

Grape training is specialised job. Therefore, it should be done by an expert in the system. Grapes are trained on different systems depending on the structure provided prior to the planting of grape vines. Some of the popular systems are discussed, under grape cultivation.

Fruit/Yield Management

1. Fruit Thinning

To improve the fruit size and quality, fruit thinning is carried out at an appropriate time.

Objectives

(*i*) Fruit thinning is necessary where the cultivar/fruit trees tend to overbear, example, peaches, grapes and plums.

(*ii*) Fruit thinning helps in the production of uniform-sized fruits required for processing and canning.

(*iii*) Thinning improves the percentage of marketable fruits.

(*iv*) The tree remains regular bearer until otherwise.

(*v*) The quality of fruit is improved, the skin colour if any get the boost.

(*vi*) The danger of limb breakage due to overbearing is reduced.

Thinning Methods

(*i*) Hand Thinning : This is a very cumbersome and laborious method. Normally the branches are shaked lightly to drop few fruits from the branch. It is advised to thin individual fruits Judiciously by hands from each branch. Keep well placed fruits with a definite ratio of leaves to the fruit, which vary species to species and cultivar to cultivar. To economise school kids or women labour should be employed for this purpose.

***(ii)* Chemical Thinning :** Some recommendations for some fruits have been made in these years. Ethrel and NAA are widely used as thinning chemicals. 2, 4, 5-T, 2, 4-D and Sevin had been tried in grape thinning. The time and concentration of chemical is very specific for a given fruit. In peach thinning ethephon @ 200-300 ppm is used 30-35 days after full-bloom or naphthalene acetic acid is applied @ 25-50 ppm 15-20 days after full bloom. NAA at higher concentrations can be harmful.

2. Spraying

To produce good quality fruits, the insect pests and diseases are to be controlled as and when these crop up. To control fungal diseases, the Bordeaux mixture sprays should be preferred over the systemic fungicides. The prevalent fungal diseases are mildew, anthracnose, blights, leaf spots, Cankers and die backs. A particular disease on a particular species/ cultivar may be serious only in a specific environment/locality, but may not be of any significance in other area. There are some diseases which appear on a fruit in all environments. It is very important to know the disease occurrence on a fruit and optimum time to spray against the disease. The help of the expert to control fungal diseases under individual fruit may be taken and required management practice adopted.

The efficacy of sprays is controlled by the type of pumps used, kind of nozzle applied for sprays. The fine sprays are always more beneficial than loose nozzles. The fungal diseases can be controlled more effectively if the fruit trees are sprayed immediately when the attack appear. The trees should be sprayed to ensure the full coverage of the foliage and branches inside the canopy.

Sprays should be made on calm days. Appropriate recommended dose should be sprayed. When controlling pest attack, sprays should only be done when attack is sufficient to reduce the quality/yield. Unnecessary sprays should be avoided. It only adds up to the management costs. No sprays may be given during harvesting of the fruit.

Micronutrient sprays should only be given at lower concentrations and in early times of leaf maturity. After the leaves become 6-7 months old they do not absorb the nutrient due to waxy lamina.

To control soil fungi in the roots of the plants. The treatments should be given immediately as the symptoms appear. Normally weak plants if bear profusely shows the attack of some disease to the roots. Smalling of leaves and early sprouting are some of the common symptoms found in all fruit plants, where roots have been infected. Regularly apply appropriate fungicide in desired quantity to each tree only then the soil fungi could be kept under control.

Fruit Harvesting and Marketing

The ultimate moto of the orchardist is to earn handsome income from his orchard. Usually orchards are auctioned or handed over to the known contractors for marketing of the fruit. The contractors market the fruit in the local markets and earn a little profit. Few contractors follow the rules of marketing to earn good profit. They pay just one half or even less to the farmer that what they actually earn. Moreover contractors do not harvest the fruit at proper maturity. While, harvesting and handling requires good deal of effort and insight. The fruit should be graded into good and poor lots to sell the fruit at a premium. Fruit should be properly packed before being sent to the market.

Harvesting Indices

For getting high profit fruits should be harvested at a desired stage of fruit maturity. Immature fruit do not develop good taste and get shrivelled very soon. On the other hand overripened fruits are damaged on the tree itself and decay starts. Most of the saleable fruit get lost in the orchard itself. The harvesting indices for sending to local or distant markets differ. Similarly the indices vary from fruit to fruit and cultivar to cultivar within a species. The climacteric fruits, examples, mango, pear and plums are harvested when fully mature but not ripe. Non climacteric fruits like citrus and grapes are only harvested when fully ripe. Many parameters are there to know the harvesting maturity of fruits. The most effective is colour of peel and T.S.S./Acid ratio. The maturity can be counted by number of days after full bloom which is not a correct method, since there are changes in the weather every year. Normally it is the experience of the grower coupled with assessing the shape colour, aroma and decrease in astringency of fruit which form the bases of fruit harvest.

Method of Harvesting

Much depends upon the method adopted for harvesting. Generally manual harvesting is carried out. It has been seen that unscrupulus contractors higher untrained labour who shake the tree and fruits fall on the ground. This is not desired. The tree roots get shaken, fruits get injured which latter get attacked by fungii. The ladders should be used to pick the fruit from higher branches. The fruits should be placed in baskets, which are piched up on the backs. The fruits can be twisted as in mango or cut with the help of sharp secatures. Litchi and guava are harvested along with a small portion of the shoot. Ber should be picked up manually without shaking the shoots. Skilled labour should be employed.

Self-marketing

The orchardist should develop the means to scare the birds and harvest the fruit at his own to market the fruit himself. Contractors earn more than the grower only working for a few months in comparison to the grower who looks after the orchard for most of the time. Moreover contractor has no love for the fruit trees. Most of the canopy is lost due to limb breakage due to the laxity at the part of the contractor. Those growers have started self-marketing have earned huge profits and continue to do so.

Fruit Handling

Harvested fruits also need careful handling. The fruits

should be placed in shade just after harvesting. The recommended size-based grading may be followed. Different grades should be packed in different packages. Wooden crates, CFB (Corrugated Fibre Board) cartoons, plastic crates, polynets and baskets prepared from the local materials are available for the use. The use of paper shavings, dry grass, rice trash, saw dust honeycomb partitions should be made to save the fruits from transport injuries.

The truck loads should be sent to distant markets. Use of air conditioned wagons can be made. For local marketing, the fruit should be transported by the available transport system locally. If the fruit is to be stored, verify the proper facility with store manager. Storage life is based on many factors like RH, storage temperature, and concentration. During storage, the senescence fruit is delayed by retarding respiratory activity of the fruit.

Marketing

The fruit fetches good price in a city market than at block or tehsil level. More profit can be have by sending the fruit to distant markets where that fruit is not locally produced. The good quality fruit of international standard can be shipped to far away countries. If will be beneficial for the marginal orchardists to form cooperatives first like in Gujrat, Maharashtra and Karnataka to save themselves from the market malpractices.

10

EFFECTS OF CLIMATE ON FRUITS AND ITS MANAGEMENT

Climate influences growth and productivity of fruit trees, much more than any other factor which affect the fruit plants. Any mistake committed at the time of planting a fruit orchard, shall affect the grower though our the life of the orchard. It can prove as mere wastage of time, energy and money. There are many agroclimatic zones in India. I.C.A.R. recognises 8 zones. National Agricultural Research Project (NARP) has further divided these zones into sub-zones on the basis of rain fall, etc. These delineations can be too many for making planning at the national level. Here the climate considering the temperature and rainfall as limiting factors for fruit growing and North Indian plains can be considered as one zone.

Mango is growing throughout the country, but its performance varies from Maharashtra to Punjab. The productivity is much lower in Punjab and Uttar Pradesh due to the frequent occurrences of frost and freezes. Grapes are performing well in south and west but in the north only one crop is taken that too is badly affected by rains. The productivity is higher in tropical region than sub-tropics. Citrus is also widely grown but its productivity varies from area to area. In Punjab and adjoining areas kinnow is very successful. Mandarins are being grown in Maharashtra. Sweet orange are mostly cultivated in arid irrigated areas of the Rajasthan and Haryana. In the Hoshiarpur belt upto Dhaula Kuan (H.P.) Kinnow is performing very well. Litchi is being successfully cultivated in Pathankot belt to Dehradun in Uttaranchal. The Almonds failed in plains of Punjab but are doing well J&K and H.P. areas.

It is difficult to form zones for fruit growing. There are many factors which comprise the climate of an area. These are discussed below with special reference to important fruits of North Indian plains which comprises Sutluj, Ganga, Alluvial plains and arid irrigated areas of Haryana and Rajasthan.

Temperature

The success and failure of the orchard is entirely dependent upon the temperature. Humidity can be supplemented by providing irrigation. Plant growth and other physiological changes, *viz.* flowering, fruiting and fruit quality are temperature dependant. In citrus the fruit maturity and quality is influenced by temperature. The temperature variations and severity of winter and frost and hot summers affect the plant growth and productivity. Temperature may fall as low as 0°C or even less in winter and may go as high as 48°C in North Indian plains. Metabolic activity is adversely affected by low and high temperatures both.

Optimum temperature varies from species to species and even cultivar to cultivar. Pear, peach and plums prefer cooler areas, whereas ber likes warm climate. Temperature determines the type of fruit crop to be planted in an area. The classification of fruit plants based on temperature requirements are Tropical, sub-tropical and temperate.

Effect of Temperature of Fruit Plants

Almost all physiological reactions in plant system are temperature dependant. Spread of diseases like scab, anthracnose, pear blight, powdery mildew are associated with temperature. physiological disorders like leaf burning, sun scalding of fruits and killing of foliage are dependent on temperature variations. The pollination carried out by insects (Bees and flies) is associated with temperature variations. It has great influence on the interval of irrigation, which affect the nutrient uptake by the feeder root system. All these actions determine the growth and development of the fruit plant. Immediate rise in temperature in April leads to fruit drop in citrus particularly kinnow and poor fruit set in soft pears.

On the basis fo climate the fruit plants have been classified as :

***(i)* Temperate :** Those fruits which require low temperature for the differentiation of fruit buds or for the completion of dormancy are known as temperate fruits. These may require high chilling hours or low for their flowering and fruiting. These fruit plants are deciduous in nature and can be grown successfully at higher elevations. Optimum temperature needed for their growth and development ranges between 25°C to 30°C. The examples are Almond, Apple, Pear, Peach, Plums, etc.

***(ii)* Tropical :** The fruit plants which require higher temperatures throughout the year, such fruit plant do not need low temperature for breaking of dormancy and for flowering. These fruit plants are generally evergreens and grow in flushes throughout the year. These plants cannot withstand too low a temperature and frost. These can only perform well where there is a little fluctuations in temperature and light. Areas between Tropic of cancer (23°.27 N latitude) and tropic of Capricorn (23°.27S latitude) come under tropical areas. The optimum temperature for their growth and development ranges between 30°C to 37°C. The fruits are mango, banana, ber, pomegranate, pineapple, etc.

Sub-tropical

These fruits need both high and low temperature for growth and fruiting. Fruits which perform better between temperate and tropical conditions are known as sub-tropical fruit plants. These may be both deciduous or evergreen and can fairly tolerate low temperature but are susceptible to frost. These fruit plants require low temperature to break the dormancy of flower buds during winter for the production of profuse flowering in the spring. These plants adjust very well to light and long dark period. Sub-tropical fruit plants need optimum temperature of 25°C to 40°C for proper growth and development. They also need low temperature up to 0°C or so without frost/freeze for fruit bud differentiation. Examples are litchi, citrus, pomegranates, low chilling requiring pears, peach and plums.

The requirements of temperature varies from species to species and cultivar to cultivar. Heat unit requirements for morphogenesis in different cultivars of pear, citrus and banana fruits varies greatly. Grape is widely adapted to temperatures of tropics and sub-tropics. The fruits produced in tropical conditions are better than sub-tropics. The same plant may require different amounts of temperature for various physiological activities. The physiological processes within a plant system may not function properly during summers when the temperature in North India reaches 47°C and winters falls as low as –1°C, due to non-availability of required amount of moisture. Due to low temperature plants may shed leaves, *e.g.* pear, peach, plum, grapes, phalsa, etc. and due to high temperature plants may go dormant, *e.g.* ber. To save fruit plants from both very high and very low temperature devices/procedures have been developed. There are nets available to check/reduce the light intensity from harsh sun. Similarly cold hardy rootstocks like Trifoliate orange can be used for citrus in cool regions. Verticle orientation of leaves on shoots can reflect sun light to save the foliage from sun burn. The fruit maturity and fruit harvesting can be regulated by using the heat unit requirements. Temperate fruit plants have more tolerance to temperature variation than tropical and sub-tropical fruit plants.

Injuries Caused by High and Low Temperature

Both high and low temperatures can cause injury to the fruit plants. These injuries may be desiccation, chilling, freezing, sun-burning and physiological injuries.

1. Desiccation

The low temperature coupled with dry air during winter may develop a situation where evergreen plants fail to tolerate the high rate of transpiration, leading to winter injury. Similar symptoms can also develop during summers. High leaf temperature due to high atmospheric temperature and humidity may lead to desiccation of leaves and branches due to less absorption of water by roots. Some examples are desiccation of leaves in Baramasi lime and young leaves of newly-planted guava plants wilt and die during hot summers. Many a times pear on kainth rootstock newly-planted plants dries up during May-June due to desiccation.

2. Chilling Injury

Tropical and sub-tropical fruit plants get injured by low temperature above freezing point. The intensity of injury depend upon the duration of the exposure to the low temperature. As we open a Refrigerator, we feel cool air touching our feet/ ground. This is because of the heavier cool air than the normal one. Similarly cool air during winters can damage the foliage and trunks upto a certain height depending upon the extent of low temperature. During 2008, the Kinnow leaves inside the plants were injured due to low temperature. The prolonged hours of low temperature can cause temporary to permanent wilting of the plant parts or whole of the tree. Papaya and guava leaves and trees got damaged due to prolonged low temperature for a month during 2008.

3. Freezing Injury

In temperate and sub-tropical climate many a times temperature falls below 0°C for long duration, resulting into formation of ice from water in the cells and tissues of the fruits plants. The tissues get killed due to enlarged size of ice crystals. The severity of injury depends upon the tolerance of the plant to freezing atmosphere. Fully grown trees of mango and guava got killed due to freeze injury during January 2008. Tree trunks of many mango and citrus plantation were damaged due to bark shreading/injury.

4. Physiological Disorders

Both higher and lower temperature than normal induces higher rate of transpiration in comparison to rate of photosynthesis leading to imbalance in the hormonal activity and metabolite synthesis and their translocation. This imbalance adversely affect various growth processes particularly apical dominance and flowering. The fruit set in soft pear and peaches is reduced due to sudden rise in temperature in March.

5. Sun-burning

High temperatures during growth period can cause death of meristematic tissues. The leaf lamina get bleached. The leaves may become devoid of chlorophyll. The fruits get injured due to sun-burning. The bark of tree trunks and scaffolds get cracked and splits. In severe cases plants get killed.

6. Sun Scald

This injury is the result of harsh sun rays which fall on the exposed sensitive parts of the fruit trees. In general citrus fruits get injured due to high temperatures in June-July. The flavedo get damaged and turn white. These fruits stop growing from injured side. Baramasi lime is the most affected due to high temperature. The fruits become lopsided and become unmarketable.

Protection from High Temperature

To protect the fruit plants from sun-injury following measures may be adopted.

1. White Washing

The newly-planted plants and young plants should be protected from sun-injury. The trunks of plants which are exposed to direct sun rays should be saved from direct sun rays by white washing the trunk portions on the onset of summers. If need be white washing may be repeated in mid-summers. White washing material can be prepared by using 10 kg. of unslaked lime in 100 litres of water +1 kg of copper sulphate.

2. Shelters

Newly-planted and young plantation can be protected from harsh weather by providing thatches (kullies) by using paddy trash or Sarkanda. The thatches should be erected 20 cm higher than height of the plants. This will provide sufficient aeration to the plants. Shade can be provided by sowing Jantar (Dhiancha) or Arhar around the plant at 1m away from the stem in basins. This protection should be sown in end February or March as the temperature warms up.

3. Wrapping of Bare Trunks

Wrap the trunks of the newly-planted plants with newspaper or farm waste materials like paddy trash, sugar cane trash. The wrapping may be strengthened by using 'seba' or synthetic string.

4. Irrigation

The damage caused by high temperature can be reduced by providing irrigation at short intervals. Drip irrigation along with mulching shall be of much use in arid and kandi areas.

5. Wind Break

Before planting an orchard wind break should be provided on the boundary by growing tall trees of Jamun, Arjuna, Poplar, Mulberry and Sisham. These tall trees along with shrubs like Jattikhati, Karonda, etc. can provide an excellent wind break. The roots of the wind break should be separated from the main plantations by digging 2 metres deep trench and 3 metres away from the wind break.

4. Low-headed Trees

The fruit plants are normally trained on modified leader system of training where lowest scaffold is selected at 45-50 cm above the ground level. The trunk is exposed to sun rays. For early years of growth, the branches are allowed to grow from the scion on the southern side at 20 cm above the ground, then these branches can provide sufficient foliage to protect the trunk from sunrays. These branches can be removed after 2-3 years later on.

Frost

Frost is the one of the major determining factor for the success or failure of an orchard in a particular locality/area. There are well marked belts where the occurrences of frost is a normal feature. Only the severity of the frost differ in different years. During 1984, January the severity of frost was great that in the Hoshiarpur belt many mango young plantations were killed outrightly. At Fruit Research Station Gangian (Dasuya) of the P.A.U. the Dusehari mango was the most damaged. Some plants of Kishan Bhog were killed. The severity of the damage on different cultivars differed to a great extent. The seedling selections of sucking type mangoes were not damaged much. The damage to foliage was only 5 to 25 percent. Among the table purpose mango cultivars the damage ranged between 15 to 50 percent. Rampur gola and Langra were the least affected. During January 2008, the frost area increased and it included Uttar Pradesh also. The plantations of aonla, guava and mango were most affected. In Punjab some mango and guava plantations were dead. Even citrus plantations suffered. The inner part foliage was damaged due to freeze. Nurseries of mango, litchi and guava plants were very badly affected.

Types of Frost

There are two types of frost one is mild frost locally called white frost and second is black frost which usually occurs in winters when there had been very little rain during the rainy season. White frost occurs when the air is inclined to be moist and dew point is below freezing point. It is a normal feature in North Indian plains. Whenever freezing temperature occurs plants are in danger of frost injury. The degree of resistance varies from species to species and cultivar to cultivar. The exact mechanism of physiological adaptation to frost is not clear. During clear and unclouded nights every surface exposed to low temperature releases heat by radiation. Loss of heat by radiation is the primary cause of frost damage. Earth loses more heat at night than it is receiving through the day in winter. Longer the supply of cold air greater the risk of white frost.

Black Frost

The drier the air lower the dew point. When there has been drought like situation prior to winter. Then the air will be more dry. When there is little moisture in the air and temperature falls below 0°C, then black frost occurs. The water in the plant tissues freezes before the dew is formed. There is no movement of air. Frost will deposit from air on any cooled surface and dew point to below freezing point. Black frost is the most dangerous. Its occurrence kills the plants.

Mechanism of Frost Damage

In plants the leaves froms the both vertical and horizontal surfaces, which are exposed to the radiation loss, because of air surrounding them get chilled to a point below which the frost starts on the leaf surfaces. Initiation of freezing in plant tissues occurs very easily when frost forms on the plant. Nucleation of freezing in tissue moisture and persistence of supercooling in the plant tissues are the factors in reaction to frost damage. Freezing results in the formation of extracellular ice, with a consequent concentration of extracellular solute, which will cause osmotic loss of water from cells.

It has been seen that after the frost damage to the foliage and apical buds, there is alcoholic smell in the whole orchard. It seems the damage caused by frost is further increased by the activity of bacterium which leads to the much damage to the plantations.

Frost Protection Measures

There can be few methods which a orchardist can adopt to check/reduce the frost damage. One can draw inferences from the nature itself. When there is cloudy night winter season there is no frost occurrence. The under surface of the clouds acts as a reflector and throws back the heat rays to earth again loss of heat by radiation is checked hence no frost. Any method which can prevent radiation loss shall be useful to check frost damage.

Similarly, wind is an other factor which can prevent the occurrence of frost. Moving wind shall stir up the air at ground level, where frost mainly occurs. Wind checks the collection of cold air in low spaces thus preserving an even temperature. Both wind and cloud can prevent frost damage to fruit plants.

Even the presence of water bodies can prevent or lower the chance of frost occurrences. In coastal areas there is no frost. Near the rivers the frost damage is much less than the faraway places. The less moist soils get more frost. Water does not produce cold air, it may provide a frictionless surface to travel for the cold air. This will help in the spread of the cold air and no or little frost. So following methods can be useful.

1. Wind Mills

These will help in the spread of cold air in the orchard and its vicinity. Mixing of hot and cold air shall prevent damage.

2. Heater/Dhooni's

Electric heaters can be installed in the orchards. There is much shortage of power in India, hence smudging can be done by burning grass or trash at places in the orchard. Fire and smoke raises atmospheric temperature in the vicinity which causes movement in hot air and then mixing of hot and cold air in the orchard, thus prevent the frost occurrence.

3. Irrigation

To apply irrigation to orchards in winter particularly during the frosty nights has been the old practice to save the plants from frost damage.

4. Covering the Young Plants

The nursery and young plantation can be provided with covers/kullies to prevent frost damage. While preparing thatches for young plants, southern side of the plant should be kept open for sun rays to enter.

5. Wrapping of Trunks

The cool air moves very near to soil level. The extra cool air during frost can damage the branches and young tree trunks so to protect these trunks, trash or gunny bags can be wrapped around the stems.

6. Spray of 'Golu'/'Pocha'/'Pandu'

The clay suspension can be sprayed before the occurrence of frost on the foliage of the fruit trees. X-ray diffraction analysis of the 'Golu' showed that it consists of smectite (50%), Illite (35%) and kaolinite (15%). The clay (Golu) suspension was sprayed as 5,10 and 20% concentrations (weight/volume) on Dusehari mango trees just before the onset of frost wave. The spray was just to wet the foliage. In this experiment it was found that unsprayed trees were damaged to 22% under sever frost occurrence. The foliage damage decreased with the increase in the Golu concentration. There was practically no damage to the foliage with 20% clay spray. The film formed by clay particles provided thermal protection to leaf lamina and apical buds by preventing loss of heat and increasing the absorption of radiant energy. When the winter season was over the clay film automatically was removed by expansion of foliage. This is a cheap and permanent method for winter months (December to March).

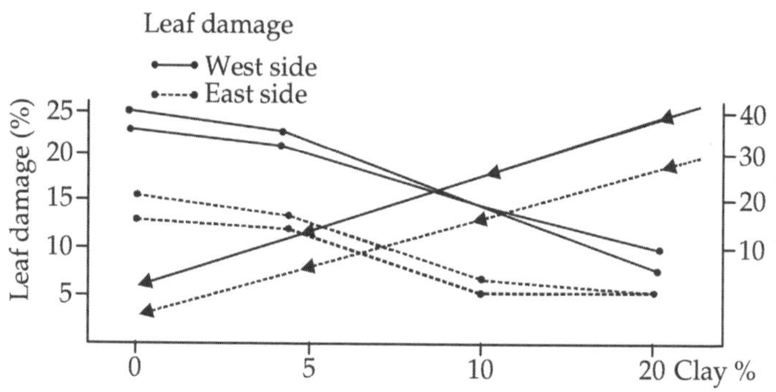

Effect of spray of clay suspension on leaf damage (%) and yield (kg/tree) of Dusehari mango

Rainfall

After temperature, rainfall is an other factor which determines the success or failure of fruit crop in an area. It affects the growth and productivity of the fruit crop. Heavy rains at the time of flowering and fruit ripening are not desired. Due to fall in water table in soils of North India, the rainfall has also declined. There has been less than optimum rains in this part of the country for the last one decade. Rains at the flowering time can wash the pollen and there can be a reduced fruit set. At fruit maturity and ripening many fungi attack the fruit causing huge loss due to fruit rotting. Well distributed rainfall help in the reduction of dependence on the irrigation water and power to lift through tubewells. Water shortage has compelled the growers to opt for drip irrigation. Rains also help in the addition of atmospheric nitrogen to the soils, besides washing the foliage of dust. The choice of fruits is dependent upon the total rainfall as well as time of rains. For example, production of grape in the north has been adversely affected by rains in June at the time of ripening of grapes.

Winds

High velocity winds both in summers and winters cause tree felling and limb breakage. The quality of fruit is also affected. The rind of fruits get damaged and bruised. The affect on soil moisture depletion necessitates frequent irrigation. The wind breaks should be provided to the orchards to prevent damage by winds.

Hails

Just like frost the hails are a normal feature in some pockets of North India. Fruit plants get damaged by hails. They affect the flowering and fruiting both. Hails at the time of fruit maturity damages the shape of fruit. The value in the market is lowered.

Light

In the Indian continent light is not a limiting factor for fruit production. Fruits and leaves exposed to light are found to improve fruit quality than unexposed. In Kinnow fruits facing light develop yellow colour earlier than the inside fruits. The TSS was more in outer fruits than those inside the tree. To get the light inside the plants little pruning is given to open the Kinnow trees to light. The intensity of light affects more than the period of light. Some fruits and foliage get sun-burnt due to harsh sun rays. In such a situation Bordeaux mixture spray can prevent this damage.

11

POST HARVEST MANAGEMENT OF FRUITS

Introduction

Many kinds of fruits are produced in India due to varied agroclimatic conditions. India itself is a good market for its own fruits. Apples and Kinnows produced in North are marketed in south and east, through out India. Similarly pineapple and banana produced in South are sold and consumed in North. Fruit cultivation is a good alternative to wheat and paddy in the north. This will save north's water resource from further falling. Total area under fruits in Punjab is 64799 hectares with annual production of total fruits 11,82,884 M.T. Punjab is largest producer of Kinnow. North India also produces maximum quantity of litchi, mango and apples. Guava and aonla are abundantly produced in Uttar Pradesh. Over 90% of Patharnakh produced in North India is supplied by Punjab alone. There is hardly any systematic arrangement for the post harvest handling of these fruits. Growers suffer over 25 percent loss of their produce only due to lack of post-harvest infrastructure and awareness of post-harvest handling practices.

There exist no proper facility for packing and grading of the harvested fruits. The cold stores available are only ment for potatoes, which provide little attention toward the storage of other vegetables and fruits. There is a dire need to increase the productivity of quality fruits per unit area along with the development of infrastructure for the post-harvest handling of fruits. Methodology needs to be standardised for the reduction of post-harvest looses, with regard to maturity standards;

ripening; packaging parameters, transportation and cold storage facilities.

Precautions Needed before Harvesting of Fruits

The growers go for orcharding only to enhance their income per unit area of their land holdings. The fruit harvest represents the end of cultural management practices in the orchard. Orchardist/contractors usually resort to unethical/ unscientific practices with respect to fruit harvesting and marketing. For example, Patharnakh pear, Kinnow and mango fruit is invariably harvested by shaking of trees. Kinnows are harvested in November when still green and acidic, but are treated with calcium carbide to develop orange colour. The scientific approach used for handling of fruit after their harvest is known as post-harvest technology. The delayed disposal of fruits to different markets from the stores cause lot of damage. Even the fruits which are marketed in time get spoiled due to various reasons. Before the fruits are harvested for storage, one must know the maturity indices for specific fruits. Maturity at fruit harvest stage is an important factor affecting the quality and behaviour of produce in post-harvest life of fruits. Equally important is the method of fruit harvest.

Indices for Maturity

It can be number of days from full bloom to fruit harvest or physical characteristics of fruit like weight and volume of fruit, firmness and colour of skin, etc. Chemical characters, *viz.*, total soluble solids, sugars, acidity and flavour of the fruit are good indices for fruit harvest.

Harvesting of Fruits

Fruits are living entity and therefore, should be harvested with great care. Fruits should be harvested with secateur or clippers and directly collected in cloth bags or crates. These should not be harvested by shaking the trees. Ground fallen fruits are likely to get injured/bruised. Fruits after harvesting should not be thrown on the orchard floor or placed in the open sun or rain.

Handling of Fruits after Harvesting

Marketing of fruits in different sizes and quality grades can fetch better price both for the producers and retailers. Some steps involved are, washing, waxing, bunching, pre-cooling and pre-packaging depending upon the kind of crop.

Grading

Grading of fruits is very essential for getting good return after their marketing. Fruits vary is size on a same tree due to many reasons. Size of fruit may vary from cultivar to cultivar, fertility level of soil and orchard management practices. Grading is carried out on the basis of the size of the fruit and quality characters.

Grading Standards for Some Fruits

Kinnow

Size Code	*Fruit Size Diameter (mm)*	*No. of Fruits in 10 kg pack*
A	60 – 64	84
B	65 – 69	72
C	70 – 72	60
D	72 – 74	54
E	75 – 79	51
F	80 – 85	45

Mango

Grade	*Fruit Weight (gm)*	*Max. Permissible Difference between Fruits within Packages (gm)*
A	100 – 200	50
B	201 – 350	75
C	351 – 550	100
D	551 – 800	125

Grapes

Grade	*Large Berries Bunch Weight (g)*	*Small Berries Bunch Weight (g)*
Extra Class	200	150
Class - I	150	100
Class - II	100	75

Guava

Size Code	*Weight (gm)*	*Diameter (mm)*
A	> 350	> 95
B	251 – 350	86 – 95
C	201 – 250	76 – 85
D	151 – 200	66 – 75
E	101 – 150	54 – 65
F	61 – 100	43 – 53

Litchi

Grade	*Fruit Diameter (mm)*
Extra Class	33
Class - I	28
Class - II	23

Pomegranate

Grade	*Fruit Weight (gm)*	*Diameter (mm)*
A	400	90
B	350	80
C	300	70
D	250	60
E	200	50

Shelf-life Enhancing Treatments

1. **Coating/Waxing :** Coating reduces the transpiration rate and improves the shelf life. Only good grade fruits should be waxed or coated to enhance shelf-life. Some

commercially available materials eq. stay fresh, citrashine are being used for this purpose successfully.

2. The pre- and post-harvest application of calcium chloride or calcium nitrate (2-4%) can improve the firmness and quality of fruits.
3. The natural colour of litchi fruits can be preserved by fumigating litchi fruits with sulphur dioxide followed by dipping in IN HCl solution. Storage of litchi at 2-3°C and with 90-95% RH can safely carried out for 3 weeks. Likewise the storage life of grapes can be improved by packing in CFB containers along with placement of grape guard paper.
4. Packing of fruits in HDPE (High density polyethylene), LDPE or shrink films.

Packaging

The fruits should be packed in corrugated fibre board boxes (CFB) or plastic crates specially designed for specific fruits, before sending to local or distant markets. The use of gunny bags or baskets of inferior material 'Shahtoot', etc. should be discouraged. These packaging materials help in easy transportation and maintain the quality of fruit during marketing and retailing. Consumers like these packagings and are willing to pay for it.

Storage

The high yielding cultivars do produce high yields by following production technologies specific for the fruit. High productivity does not mean increased income to the orchardist. High yields are normally associated with glut in the market and subsequently low rate. Fruits are perishable in nature, hence can not be with held at home or on the tree for long. So to tide over the glut period, fruits can be stored for few weeks/days. Some of the methods of storing fruits are discussed.

Cold Storage

To maintain the freshness of fruits control over temperature is very important in cool stores. Refrigerators bring down the temperature. The best condition in cold storage is the optimum temperature which should not cause chilling injury to fruits. The ideal temperature is fruit specific, hence any change from the recommended temperature shall be harmful. Relative humidity is another important factor, which must be given due attention. The fruits should be stored at a given temperature by stabilizing relative humidity (RH) in the storage room. The recommended cool temperature and RH for some of the fruits of north is given below.

Fruit	*Temperature °C*	*Relative Humidity (RH%)*	*Storage (Days)*
Kinnow	5 – 6	90 – 95	60 – 70
Oranges	4 – 6	85 – 90	180
Grape fruits	10 – 12	90	70 – 112
Guava	5 – 10	90	14 – 21
Grapes	–1	90 – 95	60 – 90
Pear	0 – 1	90 – 95	60
Peach and Plum	0 – 1	90 – 95	15 and 21
Lemon	10 – 14	90	60 – 150
Lime	9 – 10	90	30 – 60
Litchi	2 – 3	90 – 95	20 – 30
Mango	10 – 12	85 – 90	14 – 25
Ber	7 – 8	90 – 95	15

Requirements for Storage of Fruits

1. Store only high quality of fruits, free from any injury.
2. Don't pack moist fruits.
3. Avoid temperature lower than the recommended one.
4. Don't overload storage rooms.
5. Provide adequate ventilation in the store rooms.
6. Keep store rooms clean.
7. Monitor the temperature and RH regularly.

8. Don't store ethylene releasing commodity along with ethylene sensative commodity.

Controlled Atmosphere (CA)

Controlled atmosphere means a change from the normal atmospheric gas composition. When the change in specific concentrations of nitrogen, carbon dioxide and oxygen is controlled strictly, it is termed as controlled atmosphere. This modification is made by decreasing oxygen concentration or by increasing carbon dioxide concentration. The CA treatment delays fruit ripening and senescence. It also checks chilling injury to fruits and reduces the rotting incidence. If it is not properly handled, some fruits may develop off flavours and irregular ripening. The design and construction of CA stores require a precision control of the system. Some recommended concentrations of oxygen and carbon dioxide for some fruits are given below :

Fruit	O_2%	CO_2%
Grape fruit	3 – 10	5 – 10
Lemon	5 – 10	0 – 10
Mango	2 – 5	5 – 10
Orange	5 – 10	0 – 10
Papaya	2 – 5	5 – 8
Pineapple	2 – 5	5 – 10
Banana	2 – 5	2 – 5

The main principle behind CA storage is that CO_2 concentration is kept higher than O_2 levels. To remove excess of CO_2 from the room a device CO_2 scrubber is provided. The relative humidity should be maintained high. The stored boxes should be moist (wet). The floor also should have wetness with water for the storage period. This will help in maintaining the fruit size and prevent the fruits from shriveling.

Chilling Injury

Some times fruits are stored along with potatoes at subnormal temperature or much low temperature than the optimum for the fruit. Under such a condition fruits develop chilling injury. This injury is different from freezing injury. Chilling injury has been observed in kinnow, Nagpuri mandarin and mango cultivars. The symptoms of chilling injury in some fruits are as follows :

Fruit	*Symptoms*
Mango	Uneven ripening, dull and blemished skin, off flavour in pulp.
Lemons and limes	Skin pitting, discoloured peel and slow degreening
Kinnow and Nagpuri Mandarin	Discolour of peel. Browning of peel at places. Degeneration of juice vesicles.
Litchi	Skin colour becomes dull.

What is Required

1. There is a need to identify the cultivars of different fruit species for prolonged shelf-life and processing purposes.
2. Maturity standards for the harvesting of different cultivars needs to be standardized.
3. Develop appropriate protocols for grading, packing and pre-cooling etc.
4. Needs to develop infrastructural facilities for grading packing, storage, transportation and marketing.
5. Post-harvest losses at various stages needs to be assessed.
6. Needs to develop biodegradable material for fruit packaging.
7. Needs to develop value added products from indigenous fruits like Karonda and Jamun.
8. To conduct research on grading of fruits and its storage at farm itself.

9. Some cheap methodology needs to be developed to extract seeds from fruits.

10. Finished products must be marketed through established marketing system in specific Agriculture Export Zones.

12

FRUITS OF NORTH INDIAN PLAINS

MANGO

Bot. Name

Mangifera indica L.

Family

Anacardiaceae

Mango has been grown in India since long and is considered to be king of fruits. Its mention has been made in Sanskrit literature as Amra. Alexander the great found a mango garden in Indus valley in 327 B.C. Ameer Khurso saint and poet from Turkoman wrote a poem on mango as early as 1330 A.D. Akbar the great (1556-1605) planted one lac mango trees in his bagh, which was named as Lakh Bagh.

Origin

Vavilov suggested 'Indo-Burma' region as the centre of origin of mango. Folklore and religious sites of Indian people are attached to mango. Mango has attained the status of the national fruit of India. It travelled to west; south Africa and Mexico through mango stones. Genus Mangifera contains 49 species of which only 41 are valid. *Mangifera indica* to which the most of the present Indian cultivars belong is of great importance. Over one thousand varieties have been reported belonging to this genus. Some other genus found growing in India are *M. sylvatica; M. caloneura, M. foetida* and *M. caesia*. At present mango is commercially grown in Asia and European countries.

Area and Production

Mango is commercially cultivated in Andhra Pradesh, West Bengal, Karnataka, Kerala, Bihar, Uttar Pradesh, Uttara Khand, Punjab and Haryana, Maharashtra and Gujarat. Mango cultivated over an area of 2309 thousand hectares with an annual production of 12750 thousand metric tonnes. India produces quality mangoes, Alphonso is highly liked by the western countries.

In Punjab, mango is being cultivated in whole of the sub-moutane belt comprising Gurdaspur, Hoshiarpur Roop Nagar, Fatehgarh Sahib, Mohali and Patiala districts. Now its cultivation has spread to arid canal irrigated areas of northern India.

Uses

Mango leaves are fed to cattle in the shortage of fodder. Leaves are also used in various ceremonies in Hindu rituals. Mango tree has certain medicinal properties. Its wood is used for furniture making and as fuel. Fruit is a source of vitamin A and C. Mango pulp is laxative in nature and have unique nutritional value. Fruit is utilized at all stages of development in various ways ranging from chutney, pickles and curries. Ripe fruit is taken after meals. Various types of syrups, nector, jams and jelleys are prepared from the pulp/juice. The stones kernels are fed to pigs. The bark of the wood is useful in the industry.

Botany

Mango belongs to family Anacardiaceae. Fruit plants such as cashew nut (*Anacardium occidentale*) and pistachio nut (*Pistacia vera*) also belong to this family. Three species of genus Mangifera found in India are *Mangifera indica* with edible fruits, *M. sylvatica* with non-edible fruits and *M. caloneura*.

Mangifira indica ($2n = 40$). Seedling trees are big is size and can grow over 20 metre high with a same spread. Grafted trees can attain a height of 8-10 metres with a dome shaped top. Mango is evergreen with spreading branches. On road sides seedling trees have errect branches. The leaves are alternate, leathery and lanceolate in shape, with short petiole. Inflorescences in mango appear mostly terminally and rarely axillary. Flowers are small both male and hermaphrodite flowers are borne on the same inflorescence, which may be of 10-40 cm long. The stamens (4-5) of different lengths are present in a flower only one or two are fertile and rest are reduced to staminodes. Ovary is one celled, oblique and compressed. The fruit is a drupe with leathery epicarp, fleshy mesocarp (edible) and a seed with hard covering (stone) endocarp. A panicle may have few to more than 1000 flowers. The ratio of male to hermaphrodite flowers vary from 4:1 to 1:1. The ratio can vary with season, from area to area and within cultivars. In northern India Dusehari, Amrapali and Langra cultivars may have 80, 85 and 65 percent perfect flowers, respectively. Some of the sucking mango selections have only 25 to 30 percent perfect flowers.

Flowering and Fruiting

The grafted mango plants can bear inflorescence in the first year of its planting. This infloresence should be removed for enhancing vegetative growth required for canopy of the tree. Well nourished grafted plants may start bearing after 3-5 years of plantings depending upon the nature of the cultivar. In northern India flowering in mango takes place in February-March. Few plants in an orchard flower in December or January. These infloresenes can be damaged by frost, which is a normal feature in this part of the country. Flowering on the same tree can continue for a month due to differences in shoot maturity. This can be avoided by with holding irrigation to the orchard during October-November and spraying 100 ppm NAA in October and repeating the same in November. It takes 5-6 months depending upon the cultivar to mature and ripen the fruits, after flowering.

In Northern India mango give flushes of new growth from February to November. The number of these flushes depends upon many factors, particularly, cultivar, nutrition and irrigation (soil moisture). In orchards with assured irrigation facilities, there can be 5-6 flushes in comparison to 3-4 in rainfed conditions. One should take more flushes in young non-bearing plants, but flushes should be restricted up to October in bearing trees. This practice shall help the growers to get regular crop with good yield.

Flower-bud Differentiation

The bud apex become dome-shaped, broaden and get round. The scales grow and conical shape of the bud is the first sign of mango bud differentiation. Flower bud differentiation depends upon many factors such as cultivar, temperature, nutrition and growing of inter crops. For example in Bagpat area of U.P. flower bud differentiation takes place in November-December, but in Punjab it takes place in September-October.

Training of the Tree

It is very necessary to give the mango trees desired training by regular pruning, for the first 3-4 years of growth. Mango needs little annual pruning, because it grows to a dome-shaped tree in a natural way remove the lower most shoots near the ground. Allow the plant to grow as single stem up to the height of 50-60 cms. Select scaffolds on all sides of the trunk at a distance of 15 to 20 cms. No scaffold should be selected one above the other at least less than 40 cms of distance. These scaffolds should be pinched at the apices to get side branches on each scaffold. Thus there can be 8-10 branches on a tree, providing the shape of an umbrella to the tree.

Soil

Mango can be grown on a variety of soils. Poorly drained soils should be avoided. It does not perform well in soils with pH more than 7.8. Alluvial soils with good amount of organic matter and soil pH between 6.5 to 7.5 are best suited for mango orchards.

Site

Normally growers feel that mango can be planted any

where as it is a hardy plant. This is not true. It is very sensitive to both harsh summer temperature and winter frosts/freeze. The site for commercial plantation should be selected in an area where already mango orchards exist. It should not be planted in the vicinity of brick-kilns, because fumes emitted by the burning of coal are harmful to the mango fruits. A disease called black tip of mango is associated with these fumes.

Climate

Mango is a tropical fruit but is successful in sub-tropical conditions. At the time of flowering the occurrence of frost and rain is harmful. The most suitable temperature for the growth of mango is 22 – 27°C. Rains at fruit maturity are beneficial for the improvement of fruit size and quality.

Cultivars

Numerous cultivars of mango are being cultivated through out the country. There are three kinds of mango cultivars depending upon the purpose, *viz.* pickle, sucking and table types. Different mango cultivars are suited to different agro-climatic zones. Hence, while selecting a cultivar for orcharding, the fruit quality, productivity and adaptability to the area should be given prime importance. For example, Alphonso is very successful in Maharashtra and Goa areas; Dusehari in Lucknow (U.P.) and Suvarnarekha in A.P. Kishan Bhog has failed in Punjab

Time of Fruit Maturity In India

Month	*Area*
February - July	Andhra Pradesh.
April - July	Gujrat, Maharashtra, Tamil Nadu.
May - August	Bihar, Karnataka, Madhya Pradesh and West Bengal.
June - August	Uttar Pradesh, Haryana, Punjab and Rajasthan.
July - September	Himachal Pradesh and Jammu Kashmir.

due to frost. Due care should be given while choosing a cultivar for commercial use. Dusehari, Langa, Chausa, Amrapali, Rampur gola, Bambay green and Alphonso along with over sixty sucking types are performing very well at Fruit Research Station, Gangian of Punjab Agricultural University. Some other new promising xariates in India are Pusa Arunima, Pisa Surya, Ambika, CISH-M_2 etc. Some important mango cultivars are discussed.

1. Alphonso

Most important cultivar of mango having export potential. It is performing well in the Ratnagiri area of Maharashtra, and to a small extent in Gujrat and Karnataka. It has been recommended for North India too Alphonso is known by different names in different areas as Kagji, Badami, Aphur and Hapur etc.

Tree is medium, upright and spreading. Fruit is of medium in size (250 g.). It has thin skin with attractive blush on yellow ground. Flesh is firm and of excellent quality. It has good TSS : acid ratio. In North India fruit ripens in mid July. TSS ranges between 19-21%. The cultivar is prone to spongy tissue.

2. Amrapali

It is a cross between Dusehari X Neelum and has been released by IARI, New Delhi. It is a dwarf and regular bearing cultivar suited for close planting. It is being popularised for its high orchard efficiency. The fruit size is little smaller than Dusehari, but ripens later than Dusehari. Under Punjab conditions it ripens in August. Fruit has good keeping quality and fruit flavour. The TSS ranges between 18-20%.

3. Banglora (Totapuri)

It is the commercial cultivar of South. It is regular and heavy bearing cultivar. Fruit is oblong, large and necked at the base, with prominent beak. Skin thick, golden in colour, flesh firm and flat in taste. Stone is oblong and hairy : TSS varies between 15-16%.

4. Banganpali (Safeda)

It is a commercial cultivar of south particularly Andhra Pradesh. Fruits command premium price in North Indian markets due to its earlyness. Fruits remain in the market from March and to July. Trees are of medium in vigour, spreading with rounded top. Fruit size medium to large (300-450g.), beakless. Skin thin and smooth, yellow in colour, flesh firm and fibreless, good quality fruit. Stone have few hairs all over. Keeping quality is good. TSS varies between 17-18%. Under Punjab conditions fruits ripen in July.

5. Bombay Green (Malda)

It is very popular cultivar of Ganga-Jamuna plains. In Punjab it is commonly known as Malda. It is heavy bearer with medium-sized fruits of light green colour. Trees are medium to large, spreading and moderately vigorous. Fruits are beakless with round spex. Skin is medium thick flesh soft, fibreless, yellowish with TSS of 17-18%. Stone is densely covered with small hairs. Fruits ripen from May-July. In Maharashtra it ripens in May and in North India it ripens in July.

6. Dusehari (Dashehari)

One of the most popular cultivar of North India, with excellent quality and size of fruit. It is being cultivated in south India also. Trees are moderately vigorous, spreading with rounded top. Fruit is oblong with round base. Shoulders are equal and fruit is beakless. Skin is medium thick smooth, yellow, flesh firm, fibreless pleasant flavour. Taste is very sweet. Stone is medium covered with fine fibre. It is a regular bearer. Fruits ripens from June-July. TSS 19-20 percent.

7. Fazli

This cultivar originated in Bhagalpur area of Bihar. It spread to North and West Bengal due to its well-sized fruits. Tree is vigorous and spreading. Large-sized fruits with stone heaving little fibre. The fruits remain light green even at ripening. TSS is 17-18 percent. In Punjab fruits ripen in August. In Bihar it ripens in July.

8. Langra

Very important cultivar of North India after Dusehari. It originated as a chance seedling in Banaras. Tree is very vigorous and spreading. It is alternate bearer requires more planting distance due to its vigour. Heavy yielder. Fruit size is medium, light green at maturity. Very strong and pleasant flavour. Stone has fine fibre all over. In Punjab it ripens in end July. TSS 19-20 percent.

9. Rampur Gola

This cultivar originated in Rampur (U.P.). Trees are vigorous like Langra. The leaves are narrower than Langra. It is some what tolerant to frost, hence suited to Punjab conditions. Fruits can be used for pickle purpose also. Fruits are round in shape, remains light green at maturity. Skin is medium thick, flesh whitish yellow and firm. Stone small in size. Taste is good. Ripens in August. TSS of pulp 18 percent.

10. Samar Bahist Chausa (Chausa)

This is one of the best late ripening cultivar of North India. It originated is a chance seedling in Malihabad (U.P.). Tree is vigorous and spreading. It is also irregular bearer. Fruits are of medium size with equal shoulders, skin medium thick, flesh firm and fibreless. Quality of fruit very good. It ripens from July to end August. TSS of pulp 19-20 percent.

Sucking Types

A good collection of promising sucking mangoes is being maintained at Fruit Research Station Gangian of the Punjab Agricultural University. Out of sixty promising sucking mango selections some selections have been released for their cultivation. The outstanding ones are GN_2, GN_3, GN_5, GN_{12} and GN_{19} (Gangian Sindhuri). It is interesting to note that all sucking types are more or less tolerant to frost than Dusehari and other table purpose cultivars.

Planting

Before planting a mango orchard, the layout and preparation of pits may be completed at least one month in advance. It is desirable to propagate own mango plants from a desired mother-tree of a true to type cultivar. In case it is not possible, then plants may be booked with the Fruit Research Station of Gurdaspur or at Gangian (Dasuya) of the Punjab Agricultural University.

Mango should preferably be planted from August to October. It can be planted in March if regular irrigations are applied in summer months.

Planting Distance

It varies from cultivar to cultivar. Langra, Chausa and Rampur gola may be planted at a distance of 11.0 metres. Semivigorous cultivars like Dusehari and Alphanso may be planted at 9 metres apart. Amrapali can be planted at 7 metres or 7 × 3.5 m apart closely planted plants are required to be given a light pruning at the time of each harvest.

Number of plants required for one hectare are as follows:

Cultivar	*Planting Distance (m)*	*Square System*	*Hexagonal System*
Amrapali	7 × 3.5	392	448
Bombay green Dusehari	7 × 7	196	224
Alphonso	9 × 9	121	143
Fazli Langra Chausa	10 × 10	100	110
Rampur gola	11 × 11	81	90

Planting in the Field

Only well-sized healthy plants should be lifted from the nursery. Care should be taken to lift 80 percent of the feeder root and tap root system in an earth ball. The earth ball should not be too big to break during transportation. From distant nursery the plants should be brought in plastic bags of

30 × 15 cm size. Some quality soil + F.Y.M. mixture may be placed in each bag before placing the lifted earth ball in the bags. It is better than trash/grass used for wrapping the earth balls. If it is not passible to get the polythene bags, than the containers (plastic/wooden) or crates should be used to pack the plants. This will help in checking the breakage of earth balls during transportation.

Remove the packing material gentlly and place the earth ball in centre of the prepared pits. The upper surface of the earth ball should be in level with the soil in the field. The plants should not be planted too high or too low in the prepared pits. Press the sides of the newly-planted plants carefully without pressing the original earth balls. Apply light irrigation immediately after planting and level the surroundings in 'wattar' conditions.

Care of the Young Plants

Provide light irrigations at an interval of 4-7 days for a month or so. To save the grafts from being damaged due to high velocity winds, wooden stakes of which lower portions are dipped in coaltar to check the attack of white ants can be provided for the first year. Apply one litre solution to each plant by mixing chloropyriphos 10 ml/litre, after one month of planting to control white ants attack. Repeat this treatment in September for the first three years of age of the plant.

Winter/Frost Protection

Young mango plants are very susceptible to low temperature and frost injury. Frost can damage grown up trees also. In North India, frost usually occurs from December to March. Hence adequate protection from winter injury should be provided to the young plants. Thatches/kullies of Sarkanda or paddy trash should be prepared in the month of November. Keep the southern side of the 'kullies' open for aeration and sunshine. If the rainy reason had little rain then it is definite that severe frost shall occur in winter. Severe frost during January 2008 killed many mango plantations. In addition to thatches, a spray of clay (Golu/Pocha) @ 20% may be given during November before the occurence of frost. Apply irrigations to keep the soil moisture in field conditions during winter months. Smudging should be carried out by burning the dry grass/ weeds or rice trash at some places in the orchards to keep up the orchard temperature. All these measures should be taken up simultaneously to protect the mango plantations from frost/ freeze injuries.

Protection during Summers

The young plants can be killed by hot summers (loo). Grow arhar around the plants to provide desired shade. It should be grown at least one metre away from the plants. White wash the tree trunks in April or wrap the trunks with paper to avoid sun injury.

Top Working/Rejuvenation

The old inferior, unproductive seedling mango trees can be improved by top working. The benefit of established root system and well developed scaffold system can be taken by grafting with a desired cultivar. The limbs of the trees are headed back during January by keeping 30 cm long stubs. The cut ends should be smeared with Bordeaux paste/paint. Many shoots emerge on the stubs during March-April. Select out growing two shoots on each stub. Thus there should be not more than 8-12 shoots over the tree. Remove the rest of the unwanted sprouts. These shoots are side/veneer grafted during August-September. A new tree is formed. Keep only successful grafts. Continue to remove the off shoots periodically.

To rejuvenate the old/frost damaged trees, no grafting is done. Rest of the procedure is same as in top working. At the time of topworking/rejuvenation it will be beneficial if 30-40 grams of Bavistin is added to the roots in 5 litres of water to check fungal attack. Sometimes of the trees are infected with root rot fungi, the sprouts emerging on the stubs may not survive and get killed soon after, bark of the trunk get loosened and tree dries up.

Inter Cropping or Fillers

Usually mangoes are planted wide apart and have juvenile period of 4-5 years. Hence, the interspace can profitabaly be utilized for growing of crops. Select the intercrop very carefully. For the proper development of root system and canopy of the young plants, aeration and moisture is very necessary. It will be beneficial to remove weeds from the basins periodically. Intercrop should not interfere with the mango plants for nutrition, light and moisture. Intercrops can be grown for the first 4-5 years of age of the plants. Wheat can be sown by providing separate irrigation system to the mango plants. Pulses like grams and massar should be preferred. In Khariff season Moong or Arhar should be grown. Vegetable growing can be useful than intercrops. In Uttar Pradesh Sugarcane and poplar are being grown as intercrops.

Fillers are a good alternative to intercrops. Fruit plants like Plums, Peaches and Papaya can be planted as a filler in mango plantations. Mango is a slow growing fruit crop. Hence, mango itself can be used a filler, which should be removed when it starts interfering with the main plants.

Irrigation

Water is life for young plants. Light and frequent irrigations give better results than flooding after long intervals. Irrigation interval depends upon type of soil, climate and source of irrigation. The young mango plants should be kept at field capacity ('Wattar') during summers by applying irrigation at 5-7 days interval. Increase this interval slowly to 20 days as the winter starts. During rainy season avoid irrigation. If intercrops are being grown, provide separate irrigation system to provide irrigation to the mango plants during April when wheat do not require any irrigation.

Under North Indian conditions, bearing trees should applied irrigation one week before flowering and then after fruit set. During winter months, the interval of irrigation may be over 25-30 days. Proper moist conditions should be kept from April to October.

Manuring and Fertilization

It has been observed that plants supplied with high quantity of farmyard manure flourish well than fed with only inorganic/synthetic fertilizers. The requirements for various nutrients may vary in different soils for different mango cultivars. In general, plants should be given nutrition depending upon the age, canopy and yearly productivity.

Non-bearing but Juvenile

The period between planting to first bearing is known as juvenile period and stage of growth as juvenility. During this period, plants need nutrition to give out flushes at regular intervals. The emerging leaves are of coppery or light green in colour depending upon the cultivar. These leaves take 25 to 30 days to develop green colour. During this period, these leaves draw their photosynthates from the older leaves. So it is very necessary to provide adequate nutrition during the growing period. It will be appropriate to split the nitrogenous fertilizer dose into 3 or 4 parts and each applied at each flush from March to September. If the inorganic/synthetic fertilizers are mixed with farm yard manure a few days before their actual application gives added advantage in terms of availability of nutrients. Such type of mixing help in checking leaching of nutrients in light soils. Mix two kilograms of farm yard manure and 20 grams of urea per year age and apply during February, April and June each year. Thus each plant shall get 6 kg of F.Y.M. and 60 gm of urea in the first year and 12 kg F.Y.M. and 120 gm. urea in second year and so on. In the third year onward additional dose of super phosphate and Muriate of potash should be applied.

Following dose of fertilizers have given good results in farmers fields on cultivar Dusehari.

Age (Years)	*F.Y.M. (kg.)*	*Urea 46% N gm.*	*Superphosphate 16% P_2O_5 gm.*	*Muriate of Potash 60% K_2O gm.*
1 – 3	5 – 20	60 – 150	100 – 250	50 – 150
4 – 6	25 – 50	200 – 300	300 – 500	200 – 300
7 – 9	60 – 90	300 – 400	500 – 750	300 – 400
10 and above	100	400	750	400

For the first three years the dose may be split in to three equal parts and applied in February, April and June. After this apply farm yard measure plus super phosphate plus muriate potash during December or in January. Urea may be split applied in February and April. If due to any reason there is no fruiting, no urea may be given during April. Since the cost of fertilizers has gone up and efficiency of production is going down due to frost every year. It is advisable to skip super phosphate and Muriate of potash in non-bearning (off) year. Leaf analysis should form the basis of additional application of fertilizers.

Deficiency Symptoms of Essential Elements

Nutrition of mango is an important orchard management practice. Canopy development and bearing both are dependent upon nutrition. Both excess and deficiency symptoms are given below.

Nitrogen (N)

Deficiency of N causes stunting in plants. The leaves become light green, do not attain full size. The old mature leaves become yellow. The flush length is reduced. In bearing trees the number of flushes is reduced and fruit size decreases with the severity of deficiency.

Phosphorus

P deficient plants show severe stunting of growth. Leaves may fall prematurely. Branches show die back symptoms. High P causes burning of old leaves which starts from the margins and progress towards mid-rib, leading to leaf fall and then drying of branches.

Potassium (K)

K deficiency appears on semi-old leaves usually in the middle of the shoots. Purplish discolouration of leaves start with mild deficiency and intensity increases with severity of deficiency. The leaves showing necrosis progresses on the more mature leaves. No toxicity of K excess has been observed in North Indian mango orchards.

Calcium (Ca)

No typical deficiency pattern on leaves has been observed. However plants remain stunted. Additions of Superphosphate as phosphatic fertilizer takes care of calcium also. Sufficient Ca is present in superphosphate.

Magnesium (Mg)

Deficiency symptoms appear on mature leaves. Lamina start yellowing from margins and dark green bands on both sides of the mid-rib are clearly seen. The deficiency can be checked by spraying magnesium sulphate @ 2g/L of water at flush time from May to July.

Sulphur (S)

Under field conditions it is very difficult to observe S deficiency. However, sulphur deficient plants show similar behaviour as N deficient. Plants grow slowly. S deficient leaves show yellowing of young leaves and scorching on the margins of lamina. Gypsum and super phosphate contain adequate sulphur to meet the requirements of mango.

Manganese (Mn)

Mn is involved in photosynthesis and chlorophyll formation in leaves, hence deficiency symptoms appear as leaf chlorosis and vein clearing. In severe deficiency, dark-brown spots appear on the yellow portion of lamina. Leaves may fall. The deficiency can be controlled by spraying manganese sulphate @ 2 gm/L of water. Two sprays are sufficient one in April and second in June.

Zinc Deficiency (Zn)

Zinc deficiency appears on new leaves and the node length is decreased. The leaves form rosette type of tops. The tips and margins of leaves become curled. Leaves show clear intervienal chlorosis. To check the deficiency spray zinc sulphate @ 2 gm/L of water. Two sprays of zinc sulphate at flushing are sufficient.

Iron (Fe)

Iron is involved in photosynthesis being a part of chlorophyll. It is found in abundance in most soils. Its deficiency causes interveinal whitish chlorosis on young leaves in new flushes. Single spray of ferrous sulphate @ 2gm/L of water during flushing cures the Fe deficiency.

Copper (Cu)

No deficiency of copper has been reported from mango orchards. The sprays of Bordeaux mixture helps in maintaining the copper content of leaves, hence, there no need to make additional sprays.

Boron (B)

Boron deficiency causes stunting, of growth with light green foliage. The mid-rib shows browning on the ventral (under) side of leaves. Borax can be sprayed if need arises.

Chlorine (Cl)

The excess of chlorine adversely affect the foliage of young plants. Leaf scorching starts from leaf tip and progresses towards petiole. Leaves dry up and abscise. Salt-free water supply can check this toxicity.

Fruit Set

In mango fruit set is very low. Many panicles do not set any fruit. The fruit set can be improved by controlling powdery mildew in time. At full bloom spray NAA @ 2 to 3 ppm to improve fruit set.

Fruit Drop

Fruit drop in mango can be controlled by spraying 10 g of horticultural grade 2, 4-D in 500 litres of water during first week of May. NAA @ 10 ppm can be sprayed at pea stage to control drop.

Fruit Harvesting and Post-harvest Management

Mango fruits should be harvested at full maturity along with pedicles. Utmost care should be taken to avoid injury. Fruit maturity can be assessed in the orchard itself when a fruit develops little colour of a cultivar or it gets light green, the fruits are mature enough to be harvested. At this stage, the orchard may be sprayed with Bavistin @ 1 gm/L of water to control fungal attack in boxes. Fully mature but firm fruits should be picked up individually or with help of a harvester machine. The trees should not be shaken for harvesting, because the fruits on falling get injured and invite rotting fungi. The maturity time varies from region to region. In Punjab, early ripening cultivars mature in mid-June to first week of July. Late cultivars mature in August.

Grading and Packaging

After harvest fruits are placed in shade under a varandah/ store. The grading is done before packing in boxes or baskets. Fruits are graded as per weight grade A-100 to 200 gm., B 201-350 gm, C 351-550 and D-551-800 gm. The different grades are packed in wooden boxes or baskets. A basket may contain 50-100 fruits. Straw is used for packing. Wooden boxes are used for distant marketing. A box may contain 10 kg of fruit. Trash and paper is used to protect the layer of fruits from the second layer. Perforated card board are also being used. The fruits are individually packed/wrapped with tissue paper or paper shavings are used for cushioning.

Storage

Reducing the post-harvest loss is an important factor for increasing the income. Green but mature mangoes are stored in cold storages at an optimum temperature of 10-15°C. If mangoes are stored in controlled atmosphere the O_2 should be 3-7% and CO_2 5-8%. Mangoes are prone to low temperature injury. Loss of flavour and undesirable softening are major symptoms of chilling injury.

Physiological

1. Alternate Bearing

South Indian cultivars are regular in bearing. But Langra, Chausa, Rampur gola are specifically alternate bearer in North India. Dusehari and Amrapali are regular bearer. So it is advised to plant regular bearing cultivars. However, application of Paclobutrazol @ 5 gm/tree in tree basins can be helpful in checking alternate bearing. Pachlobutrazol should be applied 3-4 months before actual flowering takes place. Removing 5-10 cm of shoot along with mango fruit at the time of harvesting also encourages new growth and help in checking alternate bearing. Checking of frost injury also helps in regular bearing in frost susceptible cultivars.

2. Malformation

It is a one of the most serious disorder affecting many mango cultivars. There are two types of mango malformation. 1. Vegetative malformation is more prevalent in the nurseries. 2. Floral malformation and third type can be mixed malformation. It is a syndrome. It is thought to be caused by mites or a fungus Fusarium monileforme or due to imbalance of hormones. Sometimes numerous small leafy structures appear in place of flowers resembling a witch's broom structure. It is commonly called bunchy top. A few or all inflorescences on a tree can be malformed.

Control

1. Cut and burn the malformed inflorescences/shoots. 2. Spray 100 ppm of NAA twice, *i.e.* in first week of October and then first week of November. NAA solution should be prepared by dissolving 1:1 g (1 gm and 100mg) of 1-Naphthyl Acetic acid ($C_{12}H_{10}O_2$) 99% pure in 100 ml of absolute alcohol and slowly pouring this solution into 100 litres of water. While preparing NAA solution care must be taken that water is not to be added to the NAA dissolved in Alcohol, rather NAA dissolved in alcohol is to be poured in water, failing which NAA will get

precipitated. Continuous use of NAA sprays for a number of years can totally eliminate malformation from the orchard.

3. Jhumka

It is characterized by setting of more number of marble-sized fruitlets at the tip of the panicle. The fruitlets remain dark-green in colour. Their shape resembles that of unfertilized fruits. These fruitlets remain attached to the panicle without increasing in size. The cause of such a situation may be

(*i*) failure of pollination and fertilization due to weather or insecticide spray at full bloom.

(*ii*) Competition of new flush for photosynthates with the developing fruits.

100 ppm NAA sprays during October and November shall take care of inflorescence development. No insecticide spray should be given at the full bloom stage. Jaggery ('Gur') @ 10 percent can be sprayed at full bloom stage to attract pollinizers.

4. Spongy Tissue

It is a physiological disorder which usually affect Alphonso cultivar of mango. North Indian cultivars are not prone to this disorder. A sponge-like non-edible patch develop in the fruit flesh during ripening. Externally fruits appear normal only on cutting the fruit the disorder is detected. The affected fruits give bad odour and are not consumed. By adopting sod culture in orchards or growing legumes as cover crops or adopting mulching of basins and keeping soil moisture near field capacity during fruit development reduces the incidence of spongy tissue development in mango fruits.

Pest Management

Since, mango is evergreen tree and grows in flushes, hence prone to the attack of many insects/pests through out the year.

1. Mango Mealy Bug (*Drosicha Mangifera)*

It attacks the plants during flowering and fruiting stage

from January to April. Its males are not harmful but females lays eggs in the soil. Large number of nymphs crawl up the tree and congregate on the growing shoots and panicles. It has become a very serious pest of many fruit plants. Parthenium is acting as a host plant. Nymphs and females suck the sap from shoots and panicles and the inflorescences dry up. Unattended mango trees are seen full of mealy bugs and thus there is no fruiting at all.

Management

(*i*) Hoeing during summers (April) will kill the eggs and pupae shall be exposed to natural enemies, and heat of the sun.

(*ii*) Destroy weeds particularly parthenium (congress grass).

(*iii*) Nymphs can be prevented from crawling up to the tree trunks by applying 20 cms wide sticky band with greasy/slippery material or alkathene sheet around the trunks by upto one metre above the ground level during December. The nymphs congregating below the band/sheet are mechanically killed or sprayed with methyl parathion 50 EC @ 2 ml/L of water.

(*iv*) Soil application of toxophene @ 225 gm/tree has been very effective.

2. Mango Hoppers (*Amritodus Atkinsoni; Idioscopus Clypealis* and *Idioscopus Niveosparsus Species*)

These species of hopper are very active during February-March at the time of emergence of inflorescences. A number of nymphs and adults attack tender leaves and emerging inflorescences and suck cell sap. Due to sap sucking, the inflorescences wither away, turn brown and flowers drop off. Severely infested trees show retardation of growth. Hoppers excrete honeydew on which sooty mould develop which affect the photosynthetic activity in plants.

Management

(*i*) Avoid closer planting of mangoes.

(*ii*) Avoid flooding of orchards too often.

(*iii*) Spray twice, once in February and then in March any one of the insecticides. Sevin 50 EC (Carbaryl) 2g/L of water or Thiodan 35 EC (Endosulphan) 2 ml/L Or Malathion 50 EC 2 ml/L of water.

3. Stem Borer (*Bactocera Rufomaculata*)

Sometimes it becomes a serious pest and attacks the tree trunks. The full grown larva is stout, it makes tunnels in to the trunk under the bark and feed on internal tissue. Sometimes sap or hard bails of excreta are seen exuding from the holes.

Management

(*i*) Clean the tunnel with a hard wire and plug it with cotton soaked in kerosene oil or chloropyriphos 20 EC (50:50) in water.

4. Mango Scale

Sometimes it appears as a serious pest in some localities. Large number of scales are seen under side of leaves. Scales suck cell sap from leaves.

Management

Spray methyl parathion 50 EC @ one ml per litre of water during March and again in September. The spray should be oriented from under side of the foliage.

5. Mango Shoot Borer (*Chlumetia transversa*)

Eggs are laid on tender leaves. Freshly hatched caterpillars bore into mid-ribs of tender leaves and then bore into new shoots. Thus the upper 4-6 young leaves dry up. Young grafted plants are severely attacked. The drying up to top leaves gives an indication of the borer attack. It is prevalent in August and September in North India.

Management

At first appearance of drying top leaves spray. Thiodan 35 EC (endosulphan) @ 2 ml/L of water.

6. Bark Eating Caterpillar (*Indarbela quadrinotata)*

It is a polyphagous pest which attacks many fruit trees in this region. It is a pest of neglected orchards. Caterpillars bore into bark at crotches and make tunnels in the wood. The presence of a ribbon type formation of the dark brown excretal pellets. Pest is active through out the year.

Management

(*i*) Remove webbing to clean the tunnels with the help of a hard wire.

(*ii*) Inject kerosene oil or chloropyriphos 20 EC in water (50 : 50) in the holes with a syringe fitted on a plastic bottle. The caterpillar shall come out of the hole which should be killed.

7. Mango Fruit Fly (*Bactrocera darsalis)*

It is the most serious pest of all mango growing regions of the world. The females lay eggs just below the epidermis of the young fruit. The maggots from the eggs start feeding on the pulp thus a brown patch with resinous material is seen on the peel. Fruits start rotting and drop. Infested fruits are unfit for human consumption.

Management

(*i*) Affected fruits falling on the ground should be dumped in 3-4 feet deep pits and covered with soil.

(*ii*) Plough the field and tree basins before the emergence of inflorescences.

(*iii*) Hang traps containing 100 ml emulsion of methyl Engenol 0.1 percent. Malathion during fruit development period (April-June) under North Indian condition.

(*iv*) Give three sprays of chloropyriphos 20 EC @ 2 ml/L of water at 20 days interval starting from Ist May.

(*v*) Dip fruits in 5 percent sodium chloride solution for one hour for killing of eggs and to remove insecticide residue.

8. *Budmite* (*Eriophyes mangiferae*)

It has been reported to be associated with mango malformation. It sucks the sap from buds and cause necrosis of tender tissues. Although controlling mite has not checked the malformation but it has been considered causal organism of malformation.

Management

(*i*) Remove all malformed panicles and destroy.

(*ii*) Spray Rogor or Metasystox @ 2 ml/litre of water during summers (May-June).

9. Leaf Gall Insect (*Apsylla cistellata*)

If not checked it becomes a serious pest for all mango cultivars. Photosynthetic activity is adversely affected. Larvae feed inside the galls.

Management

(*i*) Spray chloropyriphos 20 EC 3-4 ml per litre when the galls are seen .The sprays to check fruit fly attack shall automatically manage gall insects too.

(*ii*) Four sprays of tar oil (2-3%) during egg laying period at weekly interval can reduce its incidence.

Diseases

Mango the 'King' of fruits of India is attacked by many fungal diseases. Some of major importance are as under.

1. Powdery Mildew

It is caused by the fungus *Microsphaera mangiferae*. Its

incidence is favoured by high humidity and cloudy weather during development of floral axis. A whitish powdery growth appears on inflorescences. The infected floral parts show necrotic streaks and eventually drop. Small fruits, branchlets and floral axis show die-back symptoms Flowers/fruit lets finally drop leaving a blackened axis.

Control

Spray karathane @ 1.0 gm/litre or wettable sulphur @ 2.5 gm/lite when the panicles start emerging. Repeat the spray after 20 days. Give another spray if powdery mass is seen on young fruits and their pedicles.

2. Anthracnose Die Back (*Colletotrichum gloeosporioides*)

It is also known as blossom blight. It is caused by *Colletotrichum gloeosporioides*. It causes heavy losses during rains. It affects shoots, inflorescences and fruits. Affected area develop dark brown to brownish black spots, which finally wither away. This infection is carried by the fruits to storage.

Control

(*i*) Cut and burn the shoots showing spots of canker, anthracnose and dead branches.

(*ii*) Spray Bordeaux mixture 2:2:250 thrice, once before inflorescences appear, then in April and August months.

3. Twing Die Back or Leaf Blight

It is considered to be most destructive disease of mango. It is caused by *Macrophoma mangiferae*. Minute light brown spots appear on shoots and leaves. Leaves may shrivel and fall down. The bark of affected shoots split length wise from which gum oozes out. Finally twings show die back. The fruits developing dark-brown spots may rot.

Control

(*i*) Select scion wood from disease free healthy mother plants.

(*ii*) Sterilize the grafting knife before use.

(*iii*) Cut and burn the affected branches. Apply Bordeaux paint on the cut ends.

(*iv*) Spray Bordeaux mixture 2:2:250 in April, July and August.

4. Stem Canker

It is caused by *Schizo-phyllium communae*. It causes discolouration and drying of leaves on one or more branches. The gum oozes out from infected parts. Branches may get killed. Small shell-like dirty white fruiting bodies of the fungus, with gills on the lower side appear in rows on the dead branches.

Control

Same as under Twing die back.

5. Black Tip

It is a physiological disorder caused by toxic gasses emitted by brick-kiln chimney. It is a serious problem in orchards located near brick-kiln. At first there develop a small etiolated area at the distal end of the fruit, which increases to cover the whole of the tip. Then it turns black. Infected fruits ripen prematurely and drop early.

Control

(*i*) Brick-kilns should be atleast one kilometre away from an orchard and chimney should be at least 20m high.

(*ii*) Spray 0.6% Borax thrice, before flowering, during flowering and then after fruit set.

(*iii*) Bordeaux mixture 2:2:250 should be sprayed at pea size and repeated after 20 days till fruit maturity.

CITRUS FAMILY RUTACEAE

Citrus comprises many species of economic importance. In India few species are commercially cultivated, which include grape fruit, lemons, limes, sweet oranges, and mandarins. Kinnow a hybrid between King mandarin × Willow leaf orange produced by H.B. Frost in 1915 and released in 1935 was introduced by Dr. J.C. Bakhshi at Abohar research station during 1954. This is being cultivated through out North India and even in other citrus growing states. The cultivation of citrus fruits have taken a boost due to the entrance of some private companies, who are providing their own plant material. It is a matter of time that will prove whether the new orange cultivars are a failure or success. If, this venture is successful then monoculture of Kinnow in this region shall be restricted. Efforts are also on to improve the technical know how for Blood Red and Nagpuri mandarin. Grape fruits are also being planted to meet the requirement of citrus fruits for sugar patients.

Origin

Citrus originated in South East Asia. North East region of India is the home of some citrus species *e.g.* mandarins. Nagpur Sangtra is grown on a large scale in the Vidharbha region of central India. Similarly in Assam, Brahmaputra Valley and Dibrugarh district are famous for mandarin production. Khasi mandarin is an important cultivar of Nilgiri hills. After mandarins, limes and lemons are also cultivated through out India.

Area

In India citrus is cultivated over an area of near about 923.2 thousand hectares with an estimated production of 8607.7 thousand metric tonnes. In Punjab citrus occupies 39.198 hectares with annual production of 734699 MT. Kinnow occupies 54.9% of the area under citrus.

Botany

Genus citrus belongs to family Rutaceae and sub-family Aurantioedae. While classifying citrus, Swingle has been considered to be a lumper (with 16 species only) and his student Tanaka a spliter (157 species). Hodgson (1967) classified citrus in to 36 species of commercial importance. This classification is the most accepted one.

Hybridization in citrus has led to produce a large number of intergeneric and interspecific hybrids. Most of them are used as potential rootstocks. For example, citranges (trifoliate orange X sweet orange), citrumelo (trifoliate orange X grape fruit), citrandarin (trifoliate orange X mandarin) and limequats (kumquat X West Indian lime). One of the parents trifoliate orange was choosen for its cold hardiness and dwarfing character as a rootstock.

Important tree characters of some species of commercial importance are described here.

1. *Citrus reticulata* balanc. (Sangtra)

Upright growing tree can grow upto height of 5 metres. Trees are less spreading. Leaves lanceolate in shape with narrowly-winged petioles. Flowers are of medium sized. Fruits globose in shape with loose peel. Seeds arc small with greenish cotyledons.

2. *Citrus sinensis* (L) Osbeck (Sweet Orange)

Trees are of medium size with rounded tops. Leaves medium in size bigger than reticulata. Petioles are narrowly winged. Flowers are small in size. Fruits subglobose or flattened globose. Seeds few to many, usually bold in size. The skin of fruit is tight with lines.

3. *Citrus paradisi* Macf. (Grape Fruit)

Trees are large spreading with bigger-sized foliage. Petioles are broadly winged. Flower size is also bigger than sweet oranges. Fruit is yellow coloured with thick rind. Seeds are large in size few to many per fruit, juice usually acidic.

4. *Citrus aurantifolia* Swingle (L) Lime

Tree is medium-sized with sharp and stiff numerous spines. Small leaves, with narrowly winged petioles. Flowers are small. Fruits greenish yellow, seeds small and highly polyembryonic having greenish cotyledons. The juice vesicles are also greenish in tinge.

5. Citrus limetioides tanaka (Sweet Lime)

Tree is of medium size with spreading and some what droping habit. Leaves light green in colour, with serrated margins. Big thorns present in axils of leaves. Flowers scented. Fruits medium in size globose remain green at ripening also.

Important Rootstock Species

1. *Citrus jambhiri* lush

Rough lemon (Jatti khatti). It is the most widely used rootstock in North India. It needs less water, suits arid conditions. Tree is medium in size, with sharpened strong spines. Leaves are light green in colour and of medium in size. The leaf tips are notched. Flowers small somewhat coloured. Fruits round in shape with irregular peel. On ripening peel turns light yellow. Seeds bold and many upto 32, highly polyembryonic may have 4-6 embryos.

2. *Citrus karna* Raf. Kharna Khatta (Khatta)

Tree is of medium size spreading than jatti khatti. Leaves are bigger than jatti khatti but are green in colour. Petioles are winged. Flowers coloured. Fruit medium-sized become orange coloured at ripening. Fruits have well developed papilla. Peel thick with rough surface. Seeds 30-34 rarely polyembryonic. There may be 2 or 3 embryos in few seeds.

3. *Citrus reshni* tanaka (Cleopatra)

Attractive tree with symmetrical growth. Thorns are absent. Leaves small but dark green in colour. Flowers are smaller than jattikhatti and kharna khatta. Fruits small, orange, loose skinned. Average number of seeds per fruit 14. Seeds are highly polyembryonic. Very good rootstock for commercial citrus cultivars, particularly Kinnow and sweet oranges.

Some Characters of Important Rootstocks

Name Characters	*Jatti khatti*	*karna khatta*	*Cleopatra*	*Rangpur Lime*	*Pectinifera*	*Carrizo*	*Savage*
Mature seeds per fruit	30	32	10	17	5	12	8
Immature seeds per fruit	0-1	1-2	4-7	1-2	0-1	10-11	1-2
Number of seeds per gram	12-14	6-7	12-14	18-20	7-8	4-5	4-5
Embryos per seed	4-5	2-3	5-7	2-3	4-5	4-5	4-5
% seed germination	80-90	70-75	55-60	70-80	50-60	70-80	40-50
Seedlings per seed	1.55	1.19	1.40	1.07	1.36	1.12	1.21

Soil and Climate

Citrus requires deep, well drained soils with 6.0-7.5 pH. However, by modifying cultural practices it can grow well from 5.5 to 8.5 pH. Citrus is particularly sensitive to salt concentrations in the soil solution. High EC can show early decline symptoms in citrus species. Well drained loams are better than clay soils. However, it can be grown from hilly areas to semi-arid irrigated areas. If the soil pH is near 8.5 then EC of soil should be less than 0.5 mm hos/cm for the success of citrus orchards.

Citrus performs well in sub-tropical conditions. It is being cultivated through out India. However, it can not withstand too low temperature for long period. Mandarins have adapted to sub-mountainous regions and are being cultivated in Himachal Pradesh, hilly areas of Assam and Satpura hills of Nagpur. Areas with assured irrigation and temperature range of 7-47°C are best suited for Kinnow and sweet oranges cultivation.

Cultivars

Large number of cultivars in genus citrus are being cultivated throughout India. Some of the commercial cultivars of North India are described here.

A. Mandarins or Mandarin Like Fruits

1. Kinnow (King X Willow Leaf) : This is the most important cultivar of mandarin like fruit. It is a cross between *Citrus deliciosa X Citrus nobilis.* The plants were planted first at fruit reach station Abohar. It performed well and now, it has become the number one commercial cultivar of North India. Tree is medium in growth with glossy and shining leaves. Leaves are lanceolate in shape. Tree is precocious in bearing and comes to bearing as early as in the third-year of age. Fruits are of medium in size and globose in shape. Rind is normally tight can become loose if kept upto February end on the tree. The colour of rind becomes golden yellow. Segments separate with little difficulty. Very sweet and juicy. TSS of juice varies from 12-15 percent with 2 to 1 percent acidity. Seeds per fruit varies from 10-25. Sometimes fruits may become seedless due seasonal change. It ripens in January. However, its fruits are used for juice purpose along with Musambi from November on ward.

2. Local Sangtra : The trees are errect growing with sparse folige. Fruits are smaller than Kinnow, but have short neck. The rind is loose and easy to peel, yellow in colour. The segments get separated easily. Seeds 4-5 sometimes seedless. The fruits ripen in December. T.S.S. varies from 8-10 percent with 2-3 percent acidity.

3. Khasi : It is commercially grown mandarin cultivar of Assam area and North Eastern states. There it is known as Sikkim or Kamala. Some plants in citrus orchards are mixed planted in the North also. Tree growth is just like local Sangtra. Fruits globose to ovate. Rind bright yellow to orange. The rind is smooth. The fruits may have short necks, very juicy. Juice is orange coloured. Seeds varies from 7-25.

4. Coorg : Most important cultivar of South India. It is grown on large scale in coorg area. Trees are vigorous and upright with dense foliage. Fruits medium with rind developing a bright orange colour at maturity. The fruit shape varies from oblate to globose. Rind thin easily peeled. Very juicy with many seeds ranging from 10-30.

5. Nagpur : It is very popular mandarin cultivar of India and famous in the world. It is also known as Ponkan. The trees are vigorous, with compact foliage leaves are lanceolate and usually erect. Fruits medium in size sub-globose. Rind loose develops cadmium colour at maturity. Fruits have small neck. Very juicy. Seeds vary from 5-7. Seedless fruits are also there TSS varies from 10-12 percent with 1-2% acidity.

B. Sweet Oranges

1. Musambi : Tree is of medium size with sparse foliage. Fruits medium in size and sub-globose in shape. The rind shows longitudinal furrows and a circular ring at the apex. The colour of rind at ripening turns light yellow. The juice is acidless with 10 percent TSS. Fruits ripen in November seeds varies from 20-25 per fruit. The plants raised on Pectinifera roots perform better than on jatti-khatti.

2. Pineapple : Tree is of medium size with spreading habit. Fruits medium-sized, round in shape. Rind colour yellow to orange, rind glossy and shining. Fruits are very juicy. TSS : Acid ratio well blended. Seeds varies from 10-20. Fruits ripen in December.

3. Jaffa : Tree is similar to pineapple. Fruits are bigger in size than pineapple, round in shape. Skin yellowish orange. TSS : Acid well blended. Number of seeds per fruits varies from 9-12. Fruits ripen in December. Plants raised on Cleopatra rootstock perform better than jatti-khatti.

4. Blood Red : Trees are vigorous than jaffa. Fruits are of medium-sized and round in shape. The skin develops deep orange colour at ripening in December. Flesh red when fully ripened TSS 10-13 percent, acidity 1-2 percent. The seed number per fruit ranges between 9-12. Juice reddish in colour. Plants budded on jatti-khatti show stem pitting and crease formation. Therefore cleopatra rootstock is being used to check decline in Blood Red orange.

5. Valencia Late : Tree is vigorous. Heavy bearer, fruits are of medium-size, slightly oval in shape. Skin turns golden yellow at ripening. Juice abundant with good blend of TSS and acidity. It is late ripening cultivar. The ripening time in North India is February. The seeds vary from 2-7 per fruit.

C. Grape Fruits

1. Red Blush : The tree is dwarf in nature. The fruits are of medium sized oblate in shape. Smooth rind, glossy and deep yellow with crimson patches, which become visible at ripening. Juice vesicles have deep crimson blush. Juice is less acidic with good TSS. Seeds usually get aborted. There can be 0-8 seeds per fruit. It ripens in end November.

2. Marsh Seedless : Trees are spreading and vigorous than Red blush. Fruits are medium to large in size oblate-roundish in shape skin light yellow and smooth. Juice have medium TSS and acidity. Mostly seedless due to seed abortion. It ripens in December.

3. Duncan : Tree is vigorous and speading. It bears big-sized fruits of oblate shape. Skin pale light yellow or creamy. Juice acidic and sweet with well marked bitterness. It contains very high number of seeds, may be upto 50. It ripens in November.

4. Foster : Fruits are medium to large in size with oblate shape skin pale yellow. Flesh pink and juice have well blended acidity. Taste is bitter. Seeds 40-50 per fruit. It ripens in November.

D. Lemons

1. PAU Baramasi : Tree is vigorous with dense foliage. The shoots drooping usually touch the ground. Fruits are lemon yellow, round in shape with rounded apex and tapering base. Skin is thin, fruit juicy and mostly seedless sometime containing few seeds. Bears throughout the year.

2. Eureka : Tree is semi-vigorous. Fruit oblong in shape with nippled apex. Skin-lemon yellow and sooth. Juice abundant and strongly acidic, with excellent flavour. Seeds rarely present Fruits ripen in August.

3. Punjab Galgal : Tree is vigorous, foliage light green in colour. Fruit is medium is size and oval in shape. Peel yellow smooth and glossy. Juice is very acidic (5-6 percent acidity), with 5-8 seeds per fruit. It ripens in November-December.

E. Lime

1. Kagzi : Tree is spreading and spring with dense small leaves and twings usually attacked by citrus canker. Fruits small round and thin skinned. Juice vesicles green in colour juice is very acidic (7 percent acidity).

F. Sweet Lime (Mitha)

1. Local : Tree is spreading with light green foliage. Stout thorns present on twings. Fruit medium globose to ellipsoid. Skin smooth with distinct aroma, Juice abundant, non-acidic and insipid. Seed 5-6 per fruit. Fruits ripens in September.

Plant Propagation

It is desirable to propagate ones own plant material for planting an orchard. Quality nursery plants are non-existant. No care is taken in raising nucellar rootstock. None of the private nurseries in North India is having its own rootstock mother plants. Nurseries depend upon the collection of seed from various agencies. A bottleneck type of bud union configuration has been seen in many Kinnow orchards, leading to early decline and poor quality of fruit. In rough lemon, there are many strains like jatti-khatti, jalandhari khatti, jambhiri-I and jambhiri-II. Hence one should be sure enough to have jatti khatti seeds for raising his oven rootstock. Raising of ones own planting material forms the basis of longevity and productivity of citrus orchard.

Raising of Rootstock

Jatti-khatti/cleopatra fruits should be collected from reliable source during December-January. Jatta khatti fruits can be collected in August-September and seeds got extracted and stored for sowing in February. Jatti-khatti seeds can be sown fresh in August also. After extraction in January seeds are washed in mild hot water (52°C) and sown immediately on raised beds of 2m × 1m in rows 10 cm apart. Seeds should be sown 1 cm deep 2 cm apart (seed to seed). Overlapping of seeds in lines should be avoided to achieve the benefit of polyembryony. For Musambi, Pectinifera rootstock should be raised. The seeds should be sown in August-September immediately after extraction from fruits. After seeds sowing beds may be treated with Bavistin @ 1 gm/L of water to avoid fungal attack.

The seedlings emerging from a single seed should be collected separately. Discard the bigger seedling being from zygotic origin, when the seedlings become 10-12 cms of height. The selected seedlings (nucellars) should be transplanted in the nursery beds or in polythene bags specially prepared for the purpose. Before planting rootstock seedlings in the nursery or polythene bags, the roots should be cut by 25 percent of the tap root. This will help in the development of more feeder root system. In the nursery beds, seedlings should be planted at 15 cm apart in 30 cm apart lines. Keep a distance of 60 cm after each 4 to 6 rows for budding operation.

Budding

When the seedlings become of pencil thickness then T-budding is done. The scion wood is taken from healthy true to type mother plants. T-budding is performed in August-September and February-March when sufficient cell sap flow is there in seedlings. The rootstock seedlings can also be raised in poly bags of 12 cm × 30 cm size. Plants propagated in bags are handy for transportation and planting. This limits the chance of earth ball breakage.

Transplanting

Citrus can be transplanted both in spring and in rainy season planting during hot months should be avoided. There is hardly any mortality if transplanting is done in September or October. Only healthy plants with more than 30 cm of height should be planted in the orchard. Utmost care should be taken while lifting the plants from the nursery soil. The earth ball of good size having 80 to 90 percent of root system should be lifted.

Planting Distance

Various planting distances are being given to citrus depending upon rootstock, vigour of the cultivar, climate and soil. In North India citrus is planted at 6m × 6m apart. Kinnow which is precocious in bearing is being planted at 3m × 3m; 4m × 4m; 5 × 5m and 6m × 3m. It is also being planted as a filler in mango and litchi plantations, where it is planted at 4.5 × 4.5m. However the most suitable planting distance for long term performance is 6m × 6m apart.

Care of Young Plant

Newly-planted citrus plants should be protected from hot summer temperature. The stems may be white washed in April or Arhar can be sown around the basins at a distance of one meter in end February. The young plants are susceptible to severe frost/freeze. Plants may be provided shelter for two winters. Regular irrigation protects the plants both from summer and winter injuries. No sprouts should come up on the stock portion. These sprouts may be removed periodically with hand. No branch on the scion should be selected up to height of 40-50 cm from the ground level. Dry wood and intermingling shoots should be periodically removed.

Training of Plants

The lowest scaffold should be selected at least 40-45 cm above the ground level. Regular visits to orchard during flushing should be made to select scaffolds on all directions on the main trunk. There should be 6-8 scaffolds at a distance of 20-30 cm. No scaffold should be selected one above the other at distance less than 15 cm. For getting high productivity strong tree with little pruning to open up the tree for proper sunshine and ventilation is necessary. The cut ends must be immediately treated with Bordeaux paint or Bordeaux paste.

Irrigation

It has become a debatable issue. Studies have shown that one of the major reasons for early decline of citrus is flooding and over irrigation. The servants of the owners apply irrigation through flooding to 4-10 acre without any segment. Such a practice damages the feeder root system by submerging for long period, which invite fungal attack and root rot. Normally citrus do not require much irrigation. To illustrate this point an example is sited. At Dhaula Kuan research station in H.P. a Kinnow orchard was planted in foot hills with meagre irrigation. It did not decline for 30 years. No shoot dried during these years. Kinnow also required less irrigation but due to over breaing need little more irrigation than sweet oranges. The water should be applied by modified basin system of irrigation at 7-20 days of interval during summer to winters.

Drip irrigation is showing good result but system is technical and poor after service by the installing company is becoming a major impediment in shifting to drip irrigation. The labour cost and their mistakes shall compel the growers to opt for drip irrigation in the near future. In drip irrigation, the quantity of water and interval can be adjusted according to needs of the plant. By installing tensiometers at different places in the tree basins, irrigation can be supplied very judiciously, without causing any damage to the root system.

Weed Control

Both dicot and monocot weeds infest the citrus orchards. Frequent ploughings/hoeings are costly and dangerous affair. Deep ploughing injures the feeder root system which ultimately heads for the tree decline. Hoeing should be done only once in a year to remove the weeds. Rotavator may be used to give a stirring to the soil and weeds during August-September.

Use of Weedicides

It is a cheap method of weed control in citrus orchards. Pre-emergence application of Diuron (5 kg/ha) or terbacil (4.5 kg/ha) or Atrazine (5-6 kg/ha) keep the weeds under control. Annual weeds can be controlled by spraying at their flowering stage with Gramaxone 24 WSC (paraquat) @ 6-7 ml/L of water. Similarly baru grass and woody weeds may be killed with the spray of round up Glycel 41 SL (Glyphosate) @ 10 ml/L of water. Parthenium may be killed at flowering stage by spraying common salt (NaCl) @ 20 percent.

Intercropping

In citrus intercropping can be done through out the juvenile period. Indiscriminate growing of intercrops like potato should be avoided. Leguminous crops like, peas, moong, mash, grams should be grown leaving the plant basins. If wheat is to be sown, separate irrigation system should be prepared for the plants to provide irrigation during March-April when wheat doesn't require irrigation. Chillies, Barseem and Jowar/Maize should not be grown in citrus orchards.

Fertilization

Over and inadequate fertilization is one of the reasons of citrus decline. Nutrition has become very costly input for fruit crops. Judicious use of manures and fertilizers helps to maintain productivity of citrus orchards. Right choice of fertilizers in an appropriate dose in split applications at critical growth stages of the crop results into maximum advantage. The fertilizers doses recommended are on higher side. The desirable fertilizer programme is one that provides maximum yield without affecting health of the trees. Fertilizers are becoming very costly, their use must be technically managed, avoiding any excess or deficiency citrus should not be given only inorganic fertilization. Orchards should be green manured after every 2 to 3 years. The fertilization should be adjusted on the basis of soil and leaf analysis. Leaf analysis can indicate the quality of the effective absorption of nutrients and their distribution to the different plant parts. The only drawback in leaf analysis is that analytical results at one time cannot be referred to other time. However, it is an important tool to confirm deficiency or toxicity as detected through visual observations. Many recommendations exist in literature for application of fertilizers. A mixture 90 gm each of N, P and K per tree is applied in first year of planting and gradually increased for 10 year and finally 450 g N and P and 900 g of K/tree. A manurial schedule recommended by P.A.U. is based on only N fertilization does not seem to hold good as farmers are adding phosphatic and Potash fertilizers of their own. Therefore a more workable fertilizer schedule is given under which may further be adjusted with the help of soil and leaf analysis.

Age of the Tree (Year)	*F.Y.M. (Kg.)*	*N (gm)*	*Superphosphate 16% P_2O_5 (gm)*	*Muriate of Potash 60% K_2O (gm)*
1-3	5-20	50-150	50-200	15-50
4-6	25-50	200-300	150-200	60-100
7-9	60-90	350-500	250-400	110-150
10 and above	100	500	500	150

The above doses are for Kinnow, however for oranges the P and K application may be given in alternate years. In addition to these fertilizers, need for micronutrient sprays should also be considered.

Apply farm yard manure and super phosphate plus Muriate of potash during December. Split the urea dose into three parts upto three years of age and after three to mature trees urea should be split into two parts. Apply one part in February second in April and third June to young plants. After three years of age apply urea ½ in February and second half in April. It is not the quantity of fertilizers applied which determine the growth and productivity. It is the method of application which is more important. The fertilizers should be well mixed in the basins uniformly under the whole canopy of the trees. Optimum levels of various nutrients for kinnow and sweet orange are given earlier. Accordingly fertilization may be adjusted.

	N (%)	*P (%)*	*K (%)*	*Ca (%)*	*Mg (%)*	*Zn (ppm)*	*Mn (ppm)*	*Fe (ppm)*	*Mo (ppm)*
Kinnow Vegetative Shoots	2.8	0.14	1.5	2.5	0.35	30	30	120	20
Sweet Orange	2.5	0.12	1.2	2.5	0.30	25	20	100	15

To supplement the leaf analysis, the deficiency symptoms are discussed. In view of these symptoms, growers can adjust the nutrition of their citrus orchards. Deficient nutrients can also be applied through foliage sprays. Foliar spray should not be given after August since waxy cuticle layer of leaves may not allow absorption of nutrients.

Nitrogen (N)

Nitrogen deficiency appear first on old leaves and then on young leaves. Entire tree show yellowing of leaves. New leaves become small in size and light green in colour. With acute deficiency mature leaves show mottling of green and yellow, then become entirely yellow and shed. Trees with sparse foliage and dead wood bear little fruit of poor quality. Excess of N fertilization is characterized by yellow leaf tip burn. Overgrowth, broad and succulent foliage. Deficiency of N can be overcome by spraying urea @ 1 percent.

Phosphorus (P)

In North India, P deficiency in citrus is rare. Excess of P-interferes with zinc uptake and its transport. Phosphorus is mobile hence moves from old leaves to younger ones. Older leaves become yellow just like in N deficiency. In severe deficiency, leaves show burnt appearance and may abscise. Fruits produced have thick skin with hollow central core and puffiness. High P and low N can produce small cupped and pointed leaves. Hence, a ratio of N to P in the tree is an important factor for having high quality of fruits.

Potassium (K^+)

K deficiency first appears on old leaves. The leaf margins and tip become yellow. Later on necrotic areas develop. Trees remain stunted with sparse foliage. In acute deficiency S shaped weak new lateral shoots emerge. In Kinnow, mosaic pattern of green and yellow develop on older leaves. Fruits on K-deficient trees become some what elliptical. Under acute deficiency, leaf tips scorch, gumming of twigs, finally twigs die-back. Excess of

K in citrus may produce excessively green fruits due to high acid content. The peel may be malformed to overcome K deficiency spray KNO_3 laboratory grade @ 10 gm per litre. Nutralise it with lime before spray.

Calcium (Ca^{++})

Chlorosis of terminal leaves starting from tip and margins of leaf lamina. Twig die back from tips. Calcium deficient plants produce fruits with shrivelled juice vesicles. For curing Ca deficiency, mix 1.5 kg of gypsum per tree in basins.

Magnesium (Mg^{++})

Magnesium is an essential constituent of chlorophyll and is involved in photosynthesis. Its deficiency causes bronzing or coppery leaf on mature leaves. Yellow streaks run parallel on both sides of the mid-rib. The leaf base show inverted V shaped area as dark green. Later on the deficiency symptoms appear on new leaves also. A typical situation arises, leaves fall from the heavily fruiting shoots within the tree and non fruiting shoots do not show any deficiency. The deficiency can be controlled by spraying $MgSO_4$ laboratory grade @ 2g/L of water.

Zinc (Zn^{++})

It is one of the most researched nutrients after nitrogen. Zinc regulates water relations and is a precursor for tryptophan which helps in the formation of auxin. Zinc deficiency in citrus is described by a number of names frenched, mottle leaf, smalling of leaves, rosetting, etc. Young leaves become chlorotic. In mature leaves a basal portion of the mid-rib remain green and rest of the lamina become yellow or pale green. With acute Zn deficiency leaves on the entire tree may show interveinal chlorosis and rosetting. Tree become filled with dead wood become and unproductive. Spray zinc sulphate monohydrate @ 2 gm/L of water in March and repeat in May. For soil application apply 200 gm of Agriculture grade zinc sulphate in the tree basins per tree.

Manganese (Mn^{++})

Deficiency symptoms appear first on young leaves and then on old leaves. Symptoms are similar to Fe-chlorosis with main difference with iron that veins and some portions adjacent to vein remain green. A miled mottled appearances of leaves is seen. In most of the cases trees show Fe-chlorosis with no Mn-deficiency or Mn-deficiency with no sign of Fe-deficiency the difference between Mn and Zn deficiency is that in Zn-deficiency presence of light yellow colour inthe interveinal areas. To control manganese deficiency, spray manganese sulphate @ 2g/L of water twice in March and May.

Iron (Fe^{++})

Iron is involved in photosynthesis being a part of porphyrin ring of chlorophyll and has a role in N_2-fixation. Fe-deficiency in citrus produces complete interveinal chlorosis except mid rib and lateral veins on young leaves. Leaves are reduced in size, fragile, very thin and shed early. In severe deficiency, leaves become totally yellow with a line of green in the basal part of mid-rib coupled with scorched margins and tips. Toxicity of iron is characterised by degradation of chlorophyll especially from intervienal areas with yellow lesions on the leaf surface.

Conditions Influencing Fe-Deficiency

Soils having high P, Mn, Zn or Cu. High rate of added lime and calcareous nature of soil. To control iron deficiency spray ferrous sulphate @ 2 g/L of water on new flushes, twice.

Copper (Cu^{++})

Normally copper is not deficient in citrus leaves, because citrus orchards are sprayed with Bordeaux mixture two to three times in a year. If no spray is given then copper deficiency can occur. Copper plays important role in enzymes performing respiration, water uptake and photosynthesis functions. Copper deficiency can occur where iron and manganese is high. Deficiency encourages the production of long angular twigs of

S-shaped. The foliage become dark green twisted or malformed. In severe Cu^{++} deficiency gum pockets develop under the bark and near buds during summers. Brown-stained area of hardened gum on the peel of the fruit is also visible.

Conditions Influencing Cu Deficiency

Excessive leaching due to flooding. Poor liming, high soil pH and calcareous nature of soil.

Molybdenum (Mo)

Deficiency of molybdenum produce narrow leaves with yellow spots. Brown gum or the lower leaf surface develop which turns black. Leaves show a characteristic pattern of whitish yellow and light green bands with paleish green foliage. Yellow spots uniformly distributed on the upper side of leaves is a typical symptom.

To control Mo deficiency, spray 0.5 gm ammonium molybdate per litre of water.

Boron (B)

Boron is required in the translocation of sugars and growth regulators from leaves. Deficiency symptoms appear first along the mid-rib and basal part of leaves. Terminal leaves turn necrotic, shed prematurely. Mature leaves appear brittle. Fruits develop gummy granules in the fruit albedo and become hard.

Spray 0.2 percent borax to control B deficiency.

Fruit Thinning

Citrus particularly Kinnow tends to bear heavily in early years of bearing. If Kinnow is left unthinned, it can bear over 300 fruits in the third or fourth year of age. In the 5th year the bearing may increase to 600 fruits per tree. Overbearing encourages early decline in Kinnow. Many orchards decline in the 8th year of age due to over bearing. There should be about 30 leaves for one fruit. Hence, thining is necessary to improve fruit size and to check early decline. Thinning should be manually carried out during last week of May. It is very essential for the first two years of bearing. Later on is automatically bear optimum fruit. In the 4th year of age, plants should bear 40-50 fruit and in the next year it shuld be 100-150 depending upon the health of plant. In the third bearing year the fruit number should be between 400-500.

Control of Fruit Drop

Normally fruit drop in citrus/Kinnow occurs twice one from 15th April to 15th May and second in September (pre harvest drop). The first drop occurs due to abrupt weather changes. Sometimes March becomes hot and at times too cold. This variation affects the pollination and plant growth. Sudden rise in the temperature during end April and shortage of moisture leads to fruit drop. This can be checked by keeping the soil moisture near field capacity by providing light and frequent irrigations. Spraying zinc sulphate + manganese sulphate @ 2 gm/litre mid-May helps in the checking of this drop.

Pre-harvest fruit drop commences after rainy season. This could be due to heavy new growth which competes with developing fruits for nutrition. This is called physiological fruit drop. This drop can be controlled by spraying 2, 4-D (Sodium salt of horticultural grade) @ 10 ppm as soon as the drop is observed in September. To prepare 10 ppm 2, 4-D solution, dissolve 5 gm of 2, 4-D in small quantity of absolute alcohol and mix it in 500 litres of water. The chemical should not be sprayed if cotton is being grown as intercrop. Care should be taken while spraying the orchard if any broad leaf crop is being grown in the vicinity.

The second cause of this drop can be due to the attack of some fungii. This drop is termed as Pathological fruit drop. The rot at stalk end starts and can be easily identified in the field. To control this drop spray Aureofungin @ 20 gm in 500 litres of water or Bavistin 500 g/500 L of water, or Bordeaux mixture 2:2:250. If three sprays of Bordeaux mixture are regularly given to the orchard every year from mid-June onward at 20-25 days interval, the risk of the pathological drop is eliminated.

Citrus Decline

Citrus particularly Kinnow occupies more than 50 percent of the area under citrus. Citrus decline is the only problem which discourages the citrus growers. Due to overbearing in Kinnow, the decline sets in early almost all citrus species are prone to decline, but mandarins are most susceptible. Performance of plants in early years of bearing is excellent. Later on shoots start drying, the scaffolds show oozing of gum, finally leading to bark splitting and drying of scaffold. Shoot drying can be observed from March to September. Slowly whole of the tree is dead. In such a situation orchard becomes a liability for the owner.

Early Symptoms of Decline

1. The foliage show micronutrient deficiency symptoms resembling Zn, Fe, Mn and Cu deficiencies.
2. Few leaves at the top of shoots start drying.
3. In advance stage, most of the foliage show sulphur deficiency type symptoms.
4. Bark of the shoots, scaffolds and the trunk of trees oozes gum and splits.
5. A part of tree or whole may dry up.
6. Fruit size get reduced quality is adversely affected. Granulation in fruits increases.
7. Roots get damaged.

This decline is termed as slow decline. But sometimes trees may die abruptly. This is termed as 'quick decline'. Quick decline may be due to incompatible rootstock or crease formation at the graft union or due to the attack of exocortis at the rootstock portion of trunk showing bark shreding. This can also occur due to phytophthora attack to the roots.

Factors Associated with Citrus Decline

Citrus decline can be termed as a syndrome. There is not

a single cause but is the effect of many causes, which may vary from field to field or area to area. Same of the factors are beyond the understanding of the orchardist like, soil characters, drainage, salt concentrations, water logging, frost injuries, etc. The factors, which the grower can take care of are

Rootstock Susceptibility : In citrus cultivation, rootstock has great affect in determining the success or failure of the orchard. Plants on a particular rootstock should be purchased from a reliable nursery. The nursery should maintain the rootstock and scion blocks for using it in the nursery production work. Some of the citrus rootstocks are susceptible to some of the viruses, rootrots, salt and waterlogging conditions. Private nurseries sometimes collect there rootstock seed from various sources, hence the rootstock seeds get mixed. The seedlings of some other species (not recommended) get mixed up, which become the major cause of decline. Cleopatra is the most promising rootstock for all the commercial citrus cultivars. Jatti khatti a strain of rough lemon is also good for Kinnow and other mandarins in North India.

Nutrition

This is the second major factor associated with citrus decline. Over and little nutrition both contribute towards decline. Citrus plants are very sensitive to micronutrient deficiencies. Most of the Kinnow orchards in this part of the country suffer from iron, zinc and manganese deficiencies. Heavy dependance on urea as N application causes leaching of N during rains/irrigations, causing N-deficiency. Follow the recommended schedule of fertilization. Cure the micronutrient deficiencies by giving sprays of micronutrient at flushing.

Irrigation

Mostly growers resort to flooding the orchards, without any segments. This type of irrigation may save some labour, but is very injurious to the citrus root system. The rootrot is caused by over irrigation. Jattikhatti root stock requires very little water, hence, light and at good intervals irrigation can greatly reduce the chance of citrus decline.

Intercropping

As plants occupy little space in early years of planting, growers opt for growing of intercrops to supplement their income. Excessive ploughing and growing of high nutrition requiring and water requiring intercrops results into citrus decline. Potato is one such crop which has caused decline to almost all orchards where ever these were grown. Wrong selection of crops which may shadow citrus plants for most of the time result into decline. Wheat which most of the orchardists grow in citrus can be of great harm because no irrigation is given in March-April. Hence, if wheat is grown as intercrop separate irrigation system for plants should be provided at the time of sowing of wheat.

Pests and Diseases

These factors also cause decline if not taken care of. Leaf miner, citrus psylla, aphid and lemon butterfly attack the foliage and devitalize the plants. Similarly phytophthora rot, gummosis and citrus canker cause decline. For increasing the life span of the orchards the pests and diseases should be controlled from time to time.

Sometimes the fruits of an orchard become fluffy and less juicy. This situation has been observed in Musambi and Kinnow fruits. This malady is termed as citrus fruit granulation. In this the juice vesicles become dry due to drying of juice. Fruits can be easily identified by weighing it in hands. These fruits are lighter in weight than the normal ones. The TSS and acidity also decreases. The marketability is adversely affected this malady is prominent in irrigated arid regions of North India. Musambi and Blood Red are the most affected cultivars. Kinnow has been found to be affected in isolated orchards. Studies conducted on citrus granulation show that this malady is also not associated with only one factor. Rootstock, nutrition, amount of irrigation and growth regulators are some of the factors which affect the extent of granulation.

Control Measures

Limit the frequency of irrigation and amount of water applied in each irrigation to meet the needs of the plants only. Avoid flooding of orchards. Keep a balance of macro and micro-nutrient application. Growth regulator sprays like NAA @ 200 ppm, 2, 4-D @ 15 ppm; 2, 4, 5-T @ 20 ppm or GA @ 10-25 ppm has shown some promise in checking the granulation. A change in water table and environment has reduced the extent of granulation. For example, Musambi and other oranges which were severely affected by granulation in 1974-75, are now showing little or no granulation in the same areas.

Fruit Maturity and Harvesting

Citrus fruit is non-climacteric, hence should only be harvested at full maturity, and ripened stage. Mostly farmers auction their orchards to known contractors, who harvest the fruit according to their choice. Kinnow is being harvested from November when it is quite immature and very acidic. This practice should be curbed since, acidity at this stage is harmful. It will be beneficial for the growers, if they harvest the fruit and do marketing of their own. Fruits should be harvested when these have attained the desired TSS/Acid ratio. Kinnow needs to be harvested only when it gives brix value of 10 percent and developed appropriate peel colour. Fully ripened Kinnows shows TSS/Acid ratio as 12:1 or 14:1. This situation develops in January. TSS further increases upto 15%.

Technique of Harvesting and Packing

Unscrupulous contractors resort to shaking of trees for the harvest of citrus fruits. Which is least desired. Fruits should be harvested by clipping from the tree very gently. The stem should be cut very close to the button. A small stem left at the bottom injures the other fruits while packing. The fruit should not be pulled with hands. Harvested kinnow fruits are very delicate, which should be treated as eggs. The ladders should be used to harvest the tops of the trees. After harvesting the fruits are placed under shade or in a 'varandha'. A little care while packing the citrus fruits can check spoilage.

Kinnow fruits should be packed in card board cartoons in which four dozen fruits are packed. Single layer one dozen packs are also available. If the harvested fruits are immediately placed in 100 gauge perforated polythene bags, individually and then packed in boxes in 3 to 4 layers, transportation will be without any substantial loss.

Spoilage can be minimized by dipping the fruits in 125 ppm Benlate or 500 ppm Bavistin solution for one minute, then air drying before wrapping in bags. These fruits can be stored for 60 days at 5-6°C temperature with RH of 80-90 percent. Healthy kinnow fruits if harvested in the first week of February and individually packed in 10 micron thick high density polythene bags and tieing it with rubber band or by electric realing can be kept at prevailing room temperature for a month.

Rejuvenation of Declining Citrus Orchards

The declining citrus orchards can be rejuvenated by following steps.

(*i*) Remove the dead and dried wood with sharp instruments during January before the new growth initiates. Treat the cut surfaces with Bordeaux paste. Bigger cuts should be treated with Bordeaux paint. Shreaded bark should be removed with the help of a piece of gunny bag brushing. The naked wood portions should be smeared with Bordeaux paint.

(*ii*) Spray monocrotophos 2 ml/L or curacuron @ 2 ml/L to control leaf miner, aphids and psylla.

(*iii*) Apply recommended doses of manures and fertilizers per age of the trees in the tree basins.

(*iv*) Spray zinc sulphate in April, iron sulphate a week after and manganese sulphate in May @ 2g/L of each.

(*v*) Get the leaf analysis report and continue to spray the deficient micronutrient along with the above recommended nutrition.

Plant Protection Measures

Insect Pests

Over 20 different insects/pests attack the citrus crop. Only few cause economical loss regularly. It is important to note that when an insecticide is sprayed to control a particular pest, the same shall take care of other pests of same nature attacking at that time. Growers should not worry about the number of pests attacking the crop, but for selecting an appropriate pesticide and optimum dose for immediate control. Among the pests leaf miner, psylla, white fly, aphids, lemon butterfly, bark eating cater pillars and fruit fly are the serious threat to citrus cultivation in northern India. A brief description of nature of damage and time of attack and appropriate management is given below.

1. Citrus Leaf Miner (*Phyllocnistis citrella*)

It is a serious pest of young growing plants and the nursery. It remains active through out the year. The damage is caused by a pale yellow larvae which make shinning silvery zigzag mines in new emerging leaves on each flush. The attacked leaves curl up, may dry and fall. The photosynthetic activity of the foliage is greatly reduced, which adversely affect the vigour of the plants. Leaf miner attack also encourages the development of citrus canker.

Management

For best management of the pest, spray each flush with Confidor (Imidacloprid) 17.80% SL @ ½ ml per litre of water or Curacuron (Profenofos) 50 EC @ 2 ml/L or Nuvacron 36 SL (monocrotophos) @ 2 ml/L. Hostathion 40 EC @ 2.5 ml/L or Chlorpyriphos @ 3.75 ml/L. Don't use only one pesticide. Change the pesticide for each flush for getting good response from the insecticides.

2. Citrus Psylla (*Diaphorina Citri*)

It remains active throughout the year with peak period of

infestation from March to November. Both nymphs and adults suck the cell sap from leaves, buds and young shoots. As a result the terminal shoots may wilt and dry, plant growth is affected. It also acts as a vector of greening disease, which is responsible for citrus decline. The adult psyllid is a wood coloured and can fly actively. It rests on the leaves with closed wings and hind end raised upward.

Management

Spray Rogor 30 EC (Dimethoate) @ 2.5 ml/L or Nuvacron 36 SL (Monocrotophos) @ 2 ml/L of water or Imidacloprid @ 5 ml/L. Avoid spraying if already sprayed for leaf miner.

3. Aphid (*Toxoptera, gossypii spp.*)

Aphids attack citrus during December to March months. It sucks the cell sap from tender shoots and leaves. The growth is adversely affected.

Management

Spray Nuvacron 36 SL (Monocrotophos) @ 2 ml/L or Rogor 30 EC (Dimathoate) @ 2.5 ml/L Metasystox @ 2 ml/L of water.

4. Lemon Butterfuly (*Papilio demoleus*)

It is highly destructive pest of all citrus species. The beautifully coloured larvae are seen on young leaves on terminal shoots. It eats up the leaf lamina from margins towards mid-rib. Later on it may feed on mature leaves and defoliate the whole shoot. The naked shoots on the tree are easily identified by looking at the faeces on the ground.

Management

(*i*) The best method is to kill the larvae manually as it is easy to pick up the caterpillars from the young plantations and nursery plants.

(*ii*) Incase of severe infestation spray Dursban 20 EC chloropyriphos @ 3-4 ml per litre or Ekalux 25 EC (quinalphos) @ 2-3 ml/L or thiodan 35 EC (Endosulphan) 2 ml/L of water or Sevin 50 WP (carbaryl) @ 2-3 g/litre of water.

5. Citrus White Fly (*Dialeurodes citri*) and *Citrus Black Fly* (*Aleurocanthus woglumi*)

Both nymphs and adults cause the damage by sucking cell sap from the tender foliage. The attacked leaves curl back ward and drop. Sooty mould develop on the honeydew excreted by the nymphs. The black coating interfere with the photosynthetic activity of the plants. Even fruits show black coating of sooty mould.

Management

(*i*) Spray Thiodan 35 EC (endosulphan) @ 2 ml/L or Hostathion (triazophos) 40 EC @ 2.5 ml/L or, Nuvacron 36 SL (Monocrotophos) @ 2 ml/L of water in April and September.

(*ii*) Avoid close planting and over irrigation.

6. Mites (*Eutetranychus orientalis*)

Mites become serious pest after rainy season. Leaves, flowers and fruits are damaged. The leaves have minute specks left by the feeding of mites. Leaves generally give dusty looks. In case of severe infestation, the leaves show mottling and finally drop. Skin of fruits show unpleasant patches. Marketability of fruits is greatly reduced.

Management

Spray Rogor 30 EC (Dimethoate) @ 2ml/L as soon as attack is seen in May-June.

7. Leaf Folder (*Psorosticha zizyphi*)

Both young and mature plantations are attacked from May to October. The larvae feed inside the leaves by webbing them together and start feeding from top to downwards. The plants remain stunted.

Management

Spray Nuvacron 36 SL (Monocrotophos) @ 1.5-2.0 ml/L of water or Dursban 20 EC (chloropyriphos) @ 2.5 ml/L or Ekalux 25 EC (Quinalphos) @ 2 ml/L.

8. Bark Eating Caterpillar (*Indarbela quadrinotata)*

This is a pest of neglected orchards. Caterpillars eats by making holes in the wood particularly near scaffold croches. It feeds on the bark under the cover which consists of fine pieces of wood and pellets of excreta.

Management

(*i*) Keep the orchard clean.

(*ii*) Remove the webbing from the holes with strong wire and inject chloropyriphos 50:50 solution or kerosene oil and plug the holes with cotton.

Diseases

1. Citrus Canker (*Xanthomonas citri*)

It is a very serious disease of citrus particularly limes. Kagji lime is most susceptible. Disease appear on leaves, twigs and fruits. On leaves it appears as yellowish spots. Which gradually enlarge, turn rough and brownish and become raised on both sides of the leaf. These spots are surrounded by a yellow halo. On peel of the fruit, the lesions become rough and corky. Kagji lime and grape fruits are highly susceptible.

Control

Spray 100 ppm Streptocycline in February, October and December Dissolve 10 gm streptocycline in 100 litres of water. Also add pure copper sulphate @ 2g/L. of water. Three sprays of Bordeaux mixture are also effective in the control of canker.

2. Scab (*Elsinoe fawcctti*)

Small dark, brown, rough, irregular raised lesions mostly

on the under side of the leaves appear. Twigs and fruits also get infected.

Control

Spray Bordeaux mixture 2:2:250 or 50% copper oxychloride @ 3g/L of water or Ziram 27 SC or Dithane M-45 @ 2.5 gm/L thrice from June to August at 20 days interval.

3. Gummosis (*Phytophthora palmivora*)

Foot rot or citro plithora. This disease is known by different names, *viz.* foot rot, root rot, collar rot or crown rot etc. Normally orchards with poor soil drainage, or flood irrigated are prone to the disease. Fungus causes rotting of rootlets, dropping of the blighted leaves. The infected plants flower heavily. Profuse gumming from the stem and branches. Shreading of bark at bud union configuration lengthwise in strips after drying.

Control

(*i*) Collect and burry deep in soil the diseased leaves or fruits.

(*ii*) Use cleopatra as a rootstock.

(*iii*) Avoid flood irrigation.

(*iv*) Paint the trunk portions with Bordeaux paint upto 20 cms of height.

(*v*) Spray Bordeaux mixture 2:2:250 thrice, *i.e.* in February before flowering, June and end July or give two applications of Ridomil MZ as paint (2g/100 ml of Linseed oil) to the infected portion of the trunk and drench (25g/10 litre of water/tree) the soil at the base of the tree in February-March and July-August or two sprary of Aliette 80 WP (2.5 g/L of water) in April and September.

4. Wither Tip

It is also called *die back* (anthracnose). It is caused by a fungus *Colletotrichum gloesporioides* or may be due to physiological causes. Symptoms of anthracnose appear on leaves, young shoots, and fruits. On leaves the necrotic spots appear as acervuli arranged in concentric rings. Dead parts of twigs give silvery grey appearance. Leaves shedding and twig die back are characteristic symptoms of wither tip. In severe cases the stem and infection of immature fruits results in fruit drop and show tree die back.

Control

(*i*) Follow the recommended practices for pest control and cultural operations.

(*ii*) Prune the diseased wood, leaves and fruits and destroy.

(*iii*) Spray the affected trees with Bavistin 1.0 gm/litre of water in February and Bordeaux mixture 2:2:250 in March and repeat the same in July-August.

5. Melanose or Stem and Fruit Rot

It is caused by fungus *Phomopsis citri*. It manifests as dark circular depressions and yellowish margins on leaves, branches and fruits. Later the spots become raised rough and light brown and yellow margins disappear. Sand paper texture develops on the surface of leaves and fruits.

Control

Spray 2:2:250 Bordeaux mixture during July-August and September.

6. Virus and Virus Like Diseases

1. Greening : It is caused by bacterium and manifests as stiff upright multiple twigs and buds. The leaves become small and mottled. Premature defoliation and branches show die back.

Control

(*i*) Use disease free bud wood.

(*ii*) Control citrus psylla (*Diaphorina citri*) as it is a vector for greening disease.

2. Tristeza : It is caused by a virus. It causes vein clearing in young leaves. The infected trees appear as their roots have been damaged. This is due to sieve tube necrosis at the bud union and crease formation, which check carbohydrate translocation from top to roots resulting in starvation of roots. Due to heavy bearing, trees usually exhaust. Inverse pits are seen on the bark at the bud union configuration.

Control

(*i*) Use tolerant rootstock as Cleopatra and Jatti khatti.

(*ii*) Control the insect vector (aphid) which causes its spread by following the sprays of insecticides.

3. Exocortis (Viroid) : The bark below the bud union shreads. The wood become naked at the soil level. Roots get damaged. Trees show stunting. Kharna khatta and citrangs, *viz* Troyer and carrizo are susceptible to exocortis.

Control

(*i*) Use tolerant rootstock like jatti khatti and cleopatra.

(*ii*) It spreads through cutting tools. Sterilise the tools before use in the orchard.

4. Ring Spot : It appears as yellow rings on mature leaves. Rings may be few to many per leaf. Rings may coalesce to form bigger patches. The severely infected plants show die-back and decline. The virus spreads through infected bud-wood.

Control : Use virus free bud wood for raising of nursery.

7. Citrus Nematode (*Tylenchulus semipenetrans*)

The female adult is oval and remain attached to the roots and suck cell sap as a result terminal growth is infected. The general plant vigour is adversely affected.

Control

Nursery should be raised on nematode free soil.

PEAR

Benus Pyrus

Family

Rosaceae.

Pear is one of the most important fruit crops of temperate regions. Fruits are tasty with pleasant flavour. Fruits are a rich source of protein (0.69 g) vitamins (vit. A 0.06 mg; vit. B –0.03mg) and minerals like calcium (8 mg), phosphorus (15 mg) and iron (0.5 mg) per 100 grams of pulp. Patharnakh has become the commercial fruit crop of Punjab. The area under pear is steadily increasing in North India. In Punjab, pear occupies an area of 2147 hectares with an annual production of 42940 tonnes. With the introduction of new promising semi-soft pear cultivars, the area under pear is likely to increase further.

The area under patharnakh increased in past thirty years due to the supply of quality nursery plants propagated on pear root suckers and Kainth seedlings. Now the quality plants of soft pears are being made available to the growers by Punjab Agricultural University nurseries.

Origin and History

Pyrus species are native to the Northern Hemisphere of the old world. European and West Asian species are native to Eastern Europe and South Western Asia. East and North Asian species (oriental group) are native to Eastern Asia including China, Japan and Manchuria. Patharnakh (*Pyrus pyrifolia*) (Burm. F. Nakai) originated in China from where Chinese merchants and settlers brought it to Amritsar's village Harsa Chhina during the time of Lord Kanishka (120-170 AD). From here patharnakh spread to other areas. In Himachal Pradesh and Uttar Pradesh, Patharnakh is cultivated under the name of Gola pear.

Area and Production

Area and production both increased substantially during 1980 but soon there was a glut in the market and farmers got panicy and resorted to the uprooting of pear orchards during 1985-86. At present, area under pear in Punjab is 2707 hectares. with annual production of 59992 MT.

Climate and Soil

Low chilling requiring pears are being cultivated in plains of North India. Cultivars requiring high chilling hours (900-1000) are cultivated at higher hills of Jammu and Kashmir, Himachal Pradesh and Uttara Khand.

Soil

Pear is not very fastidious to its soil requirements. It is being grown from arid irrigated areas of Sirsa-Abohar to loam and clay loam soils of Amritsar and Hoshiarpur. It is doing well in foot hills where soils are light sandy to gravels. It prefers deep well drained loam soils with pH less than 8.5. Alkaline soils are unfit for pear cultivation. The soils with electric conductivity less than 1.5 mm hos/cm, less than 10 percent $CaCO_3$ and lime 20 percent are suitable for pear cultivation. High pH soils show iron chlorosis and zinc deficiencies in the pear plants.

Climate

Pear can be grown successfully at 1200 to 1800 metre above sea level. In can tolerate very low temperature below 0°C and can with stand high temperature during summers (47°C). Best temperature for its flowering and fruiting is 2°C in winter and 32°C is summers. Rainfall 100-125 cm is sufficient for its growth. Summers should be less humid. Fruits make good growth if rains are there at maturity stage.

Botany

Pear belongs to genus Pyrus and family Rosaceae. The pear fruit is a pome. Thalamus develops and overgrows the ovary just like in apple. Ovary have 3-5 seeds. Some of the important species are discussed :

1. ***Pyrus pyrifolia*** **(Burm) F. Nakai Sand Pear :** Trees are spreading in nature and can grow upto 9 m high. The bark of the trunk is light brown, rough and shreading. Branches grow fast and become drooping. Leaves 10-12 cm long with Serrated margins, round base and accuminate apex. The upper side of the lamina is dark green and lower side of light green in colour. Fruits are round to pyriform in shape and have deciduous calyx. Fruits tastes sweet. Flesh white in colour, fruits usually hard at maturity, can with stand transportation.

2. ***Pyrus communis*** **L. (European Pear) :** Trees are of pyramidal in shape, upright growing. Bark of the trunk light brown to grey in colour. Twigs show lenticles young shoots light brown in colour. Leaves medium in size, ovate 5-10 cm long with fine serration. The base of leaves is obtuse with acuminate apex. Leaves are dark green from upper surface and light green from lower surface. Fruits variable pyriform. The lengths of neck differ from cultivar to cultivar. Fruits with fleshy pedicels and persistent calyx. Fruits sweet, soft fleshed and creamy white.

3. ***Pyrus serotina*** **Rehd. (Shiara) :** Tree is spreading. The bark of the trunks dark brown. Leaves just like *Pyrifolia* but smaller in size ovate with pointed type. The spurs are stout and spring fruits round, hard and brown in colour. Fruits have bold seeds. The taste is sour. Seeds are used as rootstock for pyrifolia and communis species.

4. ***Pyrus pashia*** **Buch-Ham (Kainth) :** Trees are spreading and upright. Tree size variable may grow 5 m to 9 m high. The bark is dark-brown. The young shoots ends in spines. The small spurs also spiny sharp. Leaves like 'patang' elliptical with fine serration on the margins and acute apex. Fruits are brown in colour hard and round in shape. It grows wild. Fruits size variable. 3-5 seeds per fruit brownish black. It is an important rootstock for pear cultivars.

Cultivars

Hard Pear (*Pyrus pyrifolia*)

1. Patharnakh : Very important cultivar of hard pear group. It is commonly known as gola pear is some areas. It is commonly cultivated cultivar in Punjab. Patharnakh requires only 250-300 chilling hours for flowering and fruiting. It is heavy bearer with well sized, firm fruits. Fruits can withstand long- distance transportation. Trees are spreading and vigorous. The scaffolds grow upright and then spread. It starts bearing at the age of 5-7 years depending upon the rootstock used. The spurs are stout and can bear fruit for 15 years, if not broken. It takes 2-3 years for taking regrowth of the spurs to bear fruit.

Fruits are of medium size, round and yellowish green with prominent dots on the surface (lenticles). Flesh white in colour, gritty due to the presence of grit cells. Flesh is crispy and juicy. Keeping quality very good, can be kept at room temperature for three to four weeks. Fruits mature in the last week of July. A fully grown tree can bear upto 3 quintals of fruit. Average yield 1.5 quintals per tree. TSS of juice 13% with 0.5% acidity.

Semi-soft Pears

(These are natural hybrids of the *Pyrus pyrifolia* × *Pyrus communis species*).

1. Punjab Beauty : The trees are upright in growth forms pyramid shape, medium vigour, regular bearer. It resembles 'Le conte' cultivar in tree shape and foliage. Fruits are medium in size, yellow in colour with red blush at full maturity. Flesh is white and juicy. TSS of juice 14% and acidity 0.3%. The fruits become soft at ripening. Fruits mature in end July. Average yield 80 kg per plant.

2. 'Le Conte' : Tree is of medium in vigour with upright and spreading habit. Fruits are of medium in size mostly pyriform in shape with small necks. The colour of the fruit is greenish yellow. Flesh white, juicy and sweet. Fruits mature first week of August. TSS of juice 14 percent and acidity 0.3%. Average yield per tree 80 kg in on year and 30 kg in off year.

3. Punjab Nectar : This cultivar is similar to the above cultivars of soft pear group. It also requires about 250-300 chilling hours. Trees grows upright and spreading at maturity. Fruits are of medium in size with 14% TSS of juice and 0.31% of acidity. Fruits ripen in end July. Average yield 80 kg per tree.

4. Punjab Gold : This is also a selection from semi-soft pear group. The young plants grow upright and are less spreading. Fruits become yellow at ripening. The TSS of juice 14.5% with 0.3% acidity. The fruits ripen in end July. Average yield 80 kg per plant.

5. Baggugosha : The trees grow similar to 'Le Conte'. The trees are taller and more spreading. It is an alternate bearer. The fruits are little smaller than 'Le conte' but are more pyriform in shape with long neck. Fruits ripen in August. Very popular in foot hills of North India. Average yield 80 kg/tree. The TSS of juice 13% with 0.3% acidity.

6. Kieffer : It is a hybrid between two species *Pyrus communis* and *Pyrus pyrifolia*. The trees are very hardy upright growing and spreading at maturity. The foliage is dense, leaves are smaller than patharnakh. Tree is very productive just like patharnakh. The spurs are stout. Fruit size is equivalent to patharnakh with prominent points at the calyx and fruits develop red bluish near maturity. The flesh is gritty. TSS of juice 12.5% with 0.3% acidity. It is late maturing cultivar. Fruits ripen in 2nd week of August. Average yield 100 kg/tree, regular bearer.

7. China : Trees are of medium vigour. It bears small-sized fruits, which ripen end June. Fruits develop red colour partially. Due to early maturity it is grown in the foot hills for early marketing. The tree yields up to 55 kg of fruit. The TSS of juice 12% with 0.32 acidity.

8. Max-red Bartlette : It is a bud sport of Bartlette. The chilling requirements are little less, that is why performs well in Punjab conditions. Tree is medium size with very high density of spurs. It is a regular bearer. Fruits ripen in August and develop red colour which makes the fruit very attractive. Average yield per tree 100 kg.

Rootstock

Many rootstocks are being used to propagate pear trees. Promising characters of rootstock are given below :

Pear Root Suckers (*Pyrus calleryana*)

This is a very old rootstock which is in use since the introduction of pear cultivar patharnakh by the chinese. Initially in rootstock trials at P.A.U. Ludhiana it has been considered to be *Pyrus pyrifolia*. A sucker was planted in the old orchard of the P.A.U. in 1976. It grew as a tree quite different from *P pyrifolia* and did not flower for twenty years, but remained vegetative. Its roots continued to give out rootsuckers, where as no rootsuckers developed on *Pyrus pyrifolia* roots. Hence it has been identified as root suckers of oriental pear *Pyrus calleryana.* Old pear orchards produce rootsuckers.

The rootsuckers are uprooted during September or October. These are planted in the nursery beds for grafting. Only the established suckers are grafted with desired cultivar during December and January. Plants get ready to transplanting in the field within a year. The rootsuckers develop a double tier root system. One, feeder roots develop in the upper 15-25 cms of soil depth and second, lower at 50-60 cms depth. Very stable rootstock for plains. The plants produced on this rootstock are semivigorous.

Pyrus pashia (Kainth)

The fruits are harvested in October. Seeds are extracted immediately and are sown directly in beds without any treatment. Seeds germinate within a week and become budable in April or may be grafted in December-January.

Kainth cuttings can also be used as a rootstock for achieving the uniformity in a rootstock. For this cuttings are prepared during December and treated with 100 ppm of IBA solution for 24 hours and then planted for rooting. Over 90% cuttings root. The shoots can be budded in April-May or grafted in December.

Kainth forms a vigorous rootstock for all the pear cultivars. Kainth roots also do not give out suckers. Its plants only produces collar suckers, which should be removed.

Pyrus serotina (Shiara)

It is very important rootstock for pear in hilly areas. Fruits are larger in size than kainth. The seeds are extracted from fruits in the same way as in kainth. The seeds are bolder than kainth. Plants propagated on this rootstock are as vigorous as in kainth. Seedlings are healthier than kainth.

Pyrus calleryana (D6)

The rootstock is commonly used in Australia. It is also a vigorous rootstock for pears. The tree growth is similar to kainth. The foliage also resembles like kainth.

Cydonia oblonga (Beedana)

This is a dwarfing rootstock for pears. In a trial at fruit research station Gangian of the P.A.U. the Patharnakh and soft pears propagated on quince died within 7-8 years of age. The bud union configuration bulged out and partition developed between scion and rootstock in xylem wood. It is compatible with high chilling requiring pear cultivars *viz*. Anjou, Old Home, Hardy and Flemish Beauty. Rootstock can be raised both through seeds and cuttings. Quince roots are resistant to pear root aplids and nematodes. Some Quince strains which are compatible to low chilling requiring soft pears should be tried.

Patharnakh (*Pyrus pyrifolia*)

Patharnakh cuttings root well. The own rooted cuttings produce precocious plants. Patharnakh form a good rootstock for soft pears. There are no rootsuckers originate from the roots. Rooting of cuttings can easily be promoted by dipping cuttings in 100 ppm IBA solution for 24 hours before planting in the beds in December. Sprouted patharnakh cuttings can be budded in April-May or grafted in December.

Own rooted plants of Patharnakh should not be planted since the vigorous trees are prone to tree felling during rainy season. Cuttings are good rootstock for semi soft pears due to semi vigorous plants.

Rejuvenation/Topworking

Old pear orchards which have ceased to bear optimum fruit can be rejuvenated. The inferior quality bearing fruit trees can be top worked with the superior scion.

Method

Head back the 4-6 well spaced scafolds to keep 30 cm long stubs during December. Completely remove rest of the branches as well as scaffolds. Apply Bordeaux paint on the cuts. Care should be taken while heading back that bark on the stubs should not get damaged. Many sprout emerge on these selected stubs, during February-March. Keep out growing two sprouts on each stub. Thus there will be 8-12 branches. The rejuvenated trees again come into bearing within 3-4 years of rejuvenation and shall bear for another 15-20 years profusely.

For topworking the sprouts on the stubs should be T-budded during April-May with an appropriate pear cultivar. The rest of the sprouts may be removed. The buds sprout within a month. Unsuccessful buds can be grafted during December. Only sprouts of new cultivar should be kept rest of the sprouts should be removed to encourage the fast growth of propagated shoots.

Selection of Plants and Planting

Prepare the pits before hand as described under layout and planting of orchards. The number required for planting on hectare of patharnakh is 169 and soft pear 272. These plants plus five percent extra should be booked with a reliable nursery in advance. Check the pear plants before hand. Plants should be straight, without any bend. The graft union should be smooth, without any plastic rings. The plants should not be older than two years and should be without spurs. The height of plants may be between one meter to one and half meter. There should be only vegetative buds present on the main axis. Insure that while lifting the plants from nursery soil all the roots have been dugged out. Head back the plants to 90 cm of height before planting in the field during December-January. Place the plants in the prepared pits so as to adjust whole root system in the pit and press from all sides. Apply irrigation soon after the actual planting.

Pear plants can also be trained in the nursery for three years near the planting site. Since pears have long juvenile period and take time for the plants fill whole of the space provided hence, these can be kept in the nursery. For this purpose, plant the pear plants at 1.5 m × 1.5 m apart. Train the plants there as per modified leader system of training. For final planting lift full root system without causing any injury to the root system and branches during December. Plant these plants inthe prepared pits so as to adjust full root system in the pits at a desired depth.

The vegetative branches may be given a little pruning so that plants may not tilt after irrigation is applied.

Care of Young Plants after Planting

Prepare the basins around the newly planted plants. Keep soil level little higher around the trunks, so that irrigation water should not stagnate near the trunks. Some plants after irrigation can tilt, straighten these at the same time. Apply 2-3 irrigations at 10 days interval. As son as temperature rises in March-April, the irrigation interval may be reduced to 7 days. During rainy season, adjust the irrigation accordingly. No irrigation is required during dormancy (December-January). It is certain that white ants will attack the young newly-planted plants. To have a check on white ants, apply chloropyriphos @ 10 ml/L of water, to the roots near the trunks. One litre of this solution may be applied to each plant at a months of interval from March to June. This practice needs to be carried out at least once every year in August or September. Increase the quantity of chloropyriphos solution to three litres at 4-5 years of age of the plant. No shoots should be allowed to sprout from the ground level to 60 cms of the trunk height. White wash the trunks up to 30 cms to save it from sun burning. Keep the pests and diseases under control.

Manuring and Irrigation

Pear need nutrition for the formation of spurs on the vegetative shoots and for flowering and fruiting later on. Trees produce heavy crops and can remove 25 kg of N, 10 kg of P, 40 kg of K and 40 kg of lime. The foliage drops during dormancy also add some nutrition.

Nitrogen should not be applied in excess as profuse vegetative growth can invite twig die back. Pear usually show iron and zinc deficiencies during summers, which should be taken care of by spraying ferrous sulphate @ 2g/L or zinc sulphate @ 2g/L respectively. The pear may be applied following doses of manures and fertilizers for getting high returns from the orchards.

Age of Tree (Years)	*F.Y.M. (Kg.)*	*Urea 46% N (gm)*	*Superphosphate 16% P_2O_5 (gm)*	*Muriate of Potash 60% K_2O (gm)*
1 – 3	10 – 20	50 – 150	100 – 300	50 – 150
4 – 6	25 – 35	200 – 300	400 – 600	200 – 300
7 – 9	40 – 50	350 – 450	700 – 900	350 – 450
10 and above	50	500	1000	500

Method of Application

It is very important to know the method of application. Growers usually hand over the fertilizers to their workers and they just spread it according to their ease near the tree trunks, which is damaging. The farm yard manure should be well rotten. Mix farm yard manure, super phosphate and muriate of potash and add by uniformly spreading in the tree basins in December-January. For the first five years of the age of the plant, the quantity of urea to be applied may be divided into three equal parts. Apply each part in April, June and August. Mix the fertilizers well with soil in the plant basins and apply light irrigation.

To the bearing plants apply farm yard manure, super phosphate and potash during December. But urea may be divided into two halves. Apply first half in February and second in April after the fruit set. The dose of this half can further be reduced if the fruit set is less. Sometimes few trees may not bear at all or bear very little fruit. In such a situation the April dose may not be applied to these trees. This situation do arise in soft pears. To get more advantage from the inorganic fertilizers, green manuring with leguminous crop should be done after every alternate year.

Deficiency of Essential Elements

Pear orchards irrigated with slug water from cities usually show deficiencies of Manganese, Magnesium, Zinc and Iron. The deficiency symptoms resemble that of citurs leaves. Leaf analysis can be got done from a nearest laboratory. However, after identification of the deficient element, the trees may be sprayed with that element. For leaf analysis collect leaf samples at the height of 2.0-2.5 metres. Take 7-8 leaves from mid shoots from all sides of the tree, totalling 50-60 leaves per tree. Wash the leaves immediately place under fan. Put these leaves in respective envelops by fully labelling, transport the samples to the laboratory.

Leaf Composition of Pear in a Healthy Orchard

N%	*P%*	*K%*	*Mn ppm*	*Zn ppm*	*Fe ppm*	*Cu ppm*
1.80 – 2.00	0.18 – 0.20	1.60 – 1.80	75	35	90	17

Patharnakh propagated through cuttings can show yellowing of leaves during summers. Such trees of on light soils show severe leaf chlorosis leading to tip burning of shoots and finally bark shreding of the branch. The chlorosis can be checked by spraying micronutrients like manganese sulphate, Ferrous sulphate, Zinc sulphate @ 2g/L of water during summers. In orchards where Bordeaux mixture is sprayed at dormancy and again in June, the copper deficiency can not exist.

Irrigation

Pear should only be grown where assured irrigation facility is available. Pear need frequent and light irrigations from March to June. Pear flowers in February and fruits grow during summer months, hence dire need of sufficient moisture. During second phase of growth of the fruits 15 June-15 July, there is huge requirement of water for the trees. If the rain is not there orchards may be irrigated by supplying sufficient irrigation. Make segments of 5-6 plants instead of making bigger segments of an acre or so. Larger segments need heavy irrigation, which may invite root rot. Saving of labour for making small segments at the cost of loosing trees is no management. Orchards which were not provided with sufficient irrigation had stunted growth of pear trees, but were free from root rot. After the fruit harvest, the interval of irrigation may be increased to be month up to November.

The making of pullies/galleries of earth on the tree trunks and scaffolds is good indication of white ant attack, which normally is seen in September and October months. Check this attack by applying chloropyriphos @ 10 ml/L. of water.

Training and Pruning of Pear Trees

It is very necessary to give proper training to the pear plants. Pear tree with strong scaffold system bear heavy crop for long periods, without any limb breakage. Pear should be trained on modified leader system of training. In the first year, no sprout should be allowed to come up on the trunks upto the height of 50-60 cms from ground level. Head back the plant at 90 cm height.

Second Year : Top most bud sprout shall become the leader. Select 3-4 outgrowing well spaced branches on the main trunk. No shoot should be one above other less than 15 cm. continue to select shoots on the leader also, by heading it back at 50-55 cms of its height. Now the total height of the plant shall be in between 1.40 m to 1.55 m. There should be 6-8 scaffolds on the whole tree. These scaffolds may be bent downward for encouraging the sprouting of a number of secondaries. Bending can be done by tieing a 'seba' with a small peg in the ground or with the trunk of the same plant. Bend the shoot gently so that shoot may not break.

Third Year : Secondary branches on the main scaffolds may be headed back to appropriate lengths, so as to promote the sprouting of buds for tertiaries

Fourth Year and Fifth Year : Unwanted secondaries/ tertiaries should be removed. Leader if not cut earlier should be headed back to appropriate length. No training is required after developing the desired scaffold system.

Pruning

After developing the scaffolds no pruning is done. Due to heavy bearing, if supports are not provided to the bearing limbs, the branches can break. Hence, during every December broken limbs/shots should be removed. Any new shoot emerging in between the main scaffolds should also be removed. While pruning care should be taken regarding pulling of cut branches to avoid spur breakage.

Weed Control

Pear orchards are infested with both annual and perennial weeds. Most common weeds are baru grass, *Dabh grass*, *Maina* and *Bathu*, some climbers, etc. Pear trees become a resting place for birds after the fruit harvest, hence some tree weeds become a great hazard, which grow underneath the pear trees.

Manual hoeing is very useful to the trees. Weeds are removed by hoeing at the time of addition of manures in December and later on. These hoeings are sufficient to check weeds in the plant basins. However, the weeds growing in the vacant places can be controlled by using chemicals.

Gramoxone is an excellent weedicide to check seeding in weeds. It can be sprayed upto tree trunks. Spray the weeds with Gramoxone (orthoparaquat dichloride 24%) @ 6-7 ml/L of water in end February or early March. Weeds shall dry up and seed shall not form. However, some weeds grow not through other means than seed are not eradicated with this chemical. For these weeds spray round up Glyphosate N (Phosphonomethyl), Glycel 41 SL. @ 10 ml/L. of water. This weedicide percolate down to the root zone and effectively kill the weeds. It should be sprayed only when weeds have attained good vegetative growth. Glyphosate should not be sprayed near the tree trunks. Better leave the basins unsprayed.

Parthenium which cannot be killed with round up should be sprayed repeatedly with Gramoxone or Spray NaCl (common salt) @ 20 percent at full growth.

To check the emergence of weeds in the vacant place spray the area with Haxuron 80 WP (Diuron) @ 8 ml/L of water. Before spray orchard may be ploughed with light tillers/ rotavator.

Growing of Inter Crops

Due to long juvenile period in pear, growing of inter crops in necessary for the grower. Only those crops should be grown which adjust to the growth behaviour of pear.

Non-bearing Orchards

Vegetables can successfully be grown. Avoid growing of potatoes. Pulses like moong, mash, massar grams and peas can also be grown. Do not grow fodders like Jowar, Maize or Berseem. Some plants, *e.g.* peach, plum can be planted as fillers in pear orchards.

Bearing Orchards

Growing of Rabi season crops should be avoided. Vegetables as radish, carrots and cauliflowers can be grown. Toria is a good crop which adjust well with pear.

Fruit Harvesting

Pear should be harvested at full maturity. Immature fruits on harvesting get shrivelled. Patharnakh takes 145 days to mature from full bloom. Soft pears take 135 days to mature. At maturity green colour of the fruit changes to light yellowish green, firmness decreases. TSS of the juice reaches 9-10 percent. Optimum harvesting time for Patharnakh in North India is end July. Similarly for soft pears it is August.

Care should be taken while harvesting the fruit. Gently pick the fruit upward, give a little twist and thumb press the pedicel at the point of attachment with the spur. Spurs should not be injured during fruit harvesting. A spur can bear fruit for 10-15 years. The broken spurs do not bear fruit for 3-4 years. Fruit pedicels also should not break from the centre. Either there should be full stalk or there should be no stalk with the fruit. Harvesting should be done with the help of ladders. Trees should not be shaken for fruit harvesting.

Grading and Packing

Harvested fruits should be kept under shade in an airy 'Varandah.' Fruits should not be wet at the time of packing. For sending the fruit to a distant market it should be packed properly in wooden boxes. A box may contain 17-18 kg of fruit. Place the fruits in layers in the boxes by placing some grass/rice trash at the bottom and at the top and cover it with paper before closing the box with its lid. Packed boxes are transported by trucks. Grade the fruits before packing.

Marketing

Different grades of the fruits are auctioned separately in the big markets. However, in the local market contractors sell fruits in gunny bags or in open crates.

Some intelligent owners market their own crop themselves. They earn handsome income through this venture. They just earn double what contractor offer to them.

Post-harvest Handling

Properly graded and packed fruits in wooden or card board boxes can be stored at 0 to 3°C with RH 85 to 90 percent. Pear can be stored in controlled atmosphere at 0-1°C with 1-2% O_2 and 0.5% CO_2 for two months. A practice adopted by middle men at Delhi market is worth to mention. They send pear boxes alternately packed with mango boxes in trucks. They say both mango and pear get benefited. Pear develops a very good colour by absording ethylene produced by mangoes, and mango do not rot. Pears should not be stored for long periods to save their competition with early apple coming to market. Pears should be stored for 20-30 days to tide over the glut period.

PLANT PROTECTION MEASURES

A. Insect-Pests

1. Aphids

Myzes and *Branchycaudus* sp. Aphids usually attack the opening buds, emerging leaves and flower pedicles. Damage is caused by sucking cell sap from tender leaves, shoots, flowers and young fruits. If left uncontrolled can cause huge damage to be foliage and fruits. Flowers and fruits can drop. Sooty mould develop on leaves and fruits. Photosynthetic activity get hindered.

Management

A dormant spray of monocrotophos 36 SL @ 1 ml/L or Rogor 30 EC (Dimethoate) 2 ml/L. just before flowering and sprouting of buds shall be very beneficial in the control of aphid. Second spray should be given after petal fall.

2. Hairy Caterpillar (*Euproctis* Sp.)

The caterpillars feed on leaves and then attack the young growing fruits usually in April. Whole of the green epidermis of the fruit is eaten up, fruits turn black and fall.

Management

(*i*) The young caterpillars can be manually killed as they collect on the under side of leaves or crawl down on the tree trunks at night. Kill these caterpillars before sunrise early in the morning.

(*ii*) Spray the trees with Thiodan 35 EC (Endosulphan) @ 2 ml/L or Dursban 20 EC (Chloropyriphos) @ 3ml/L. or Sevin 50 WP (carbaryl) 2g/L of water.

3. White Ants

Tender roots of newly-planted pear plants are very much liked by the ants. Ants may eat up whole of the xylem of the root and bark of the plant. The leaves turns yellow. Plant may dry up within a few days.

Management

Apply a litre of solution of chloropyriphos @ 10 ml/litre of water after a month of newly planted plants. Repeat the treatment if need be as the age of the plant increases, increase the volume of solution to 2-3 litres.

4. Plum Case Worm (*Cremastopsychae pendula*)

It nibbles on the bark tender shoots and branches. It feeds on leaves and make holes in the lamina. It also feed on the epidermis of young fruits. Quality of fruits deteriorates.

Management

As under hairy caterpillars.

5. Bark Eating Caterpillar (*Indarbela quardinotata*)

Caterpillar make holes in to stem, particularly in croches/ joints. It eats under the cover of webbing containing excreta and wood.

Management

Remove excreta with the help of wire and inject kerosene oil or chloropyriphos 50:50 water solution as soon as the attack is seen. If left unattended whole orchard get infected and become difficult to manage later on.

6. Mites (*Eutetranychus orientalis*)

It appear on leaves in summers (May-June). Mites suck the cell sap from leaves and fruits. Brownish spots appear on leaves. If incidence is high leaves become pale, dryup and fall down. Whole of the tree give dusty look.

Management

Spray Rogor 30 EC (Dimethoate) @ 2 ml/L of water or Fosmite 50 Ec (ethion) @ 1 ml/L.

7. Fruit Fly

Fruit fly punctures the fruit from maturity to ripening stage. Fruits rot at the place of puctures. The juice oozes out. Fruits are unfit for consumption.

Management

(*i*) Dump the damaged fruits in deep pits. Keep the orchard clear of dropped fruits.

(*ii*) Spray Dursban 20 EC (Chloropyriphos) @ 4 ml/L of water in the first week of June.

(*iii*) Spray Thiodan 35 EC (Endosulphan) @ 2 ml/L of water after 20 days of chloropyriphos spray.

B. Diseases

1. Pear Scab (*Venturia pyrina*)

Olive dark mouldy spots appear especially on the under side of the leaf. Later on the spots turn grey and necrotic. Similar spots develop on the flower pedicels, calyx and petals. The affected flowers usually drop. Dark-brown spots are produced on the fruits. The spots join together become corky and crack.

Control

(*i*) Use bulky organic manures (farm yard manure) double the quantity recommended per tree.

(*ii*) Spray Captan 2g/litre before blossom starts appearing. Repeat the spray after full bloom. If disease persists give more sprays.

(*iii*) Collect and burn or carefully plough the fallen leaves to minimize ascospore discharge and consequent infection. If trees have intermingled with each other, open the trees to promote aeration by giving light pruning during December.

2. Shoot/Fruit Blight and Bark Canker (*Phoma glomerata*)

Cankers appear on budscars, wounds on twigs, stubs or in crotches. Cankers are elliptical in shape. Small circular brown spots appear around a leaf-scar or superficial wound. As the canker enlarges the centres become sunken with the surrounding healthy bark. The fungus perpetuates itself on the trees in bark cankers. The apical shoots become leafless and dry up.

Control

(*i*) Cankers on the trunk and in the crotches should be removed and the dead bark decorticated along with 2 cm of the healthy bark. All the dead wood and prunings should be destroyed.

(*ii*) Cover the wounds with disinfectant solution and apply Bordeaux paste. After one week cover the paste with Bordeaux paint. Spray Bordeaux mixture 2:2:250 to the dormant trees and repeat after fruit set and in August after fruit harvest.

3. Root Rot and Sapwood Rot

It is caused by *Polyporus palustris*, *Ganoderma lucidum* and *Schizophyllum communi*. Effect on main trunk, limbs and secondary branches can be seen. The bark and wood of the roots rot and turn brown, with white mycelial mat in the crevices. The affected trees show smalling of leaves, heavy flowering and fruit set. The leaves become spars. Finally tree looks sickly and dries up. It can fall before drying due to load of the fruits.

Control

(*i*) Locate the affected trees, showing cuping of leaves, spars foliage, sickly appearances, overfruiting and profuse and early flowering symptoms.

(*ii*) Apply 10 grams of Bavistin 50% WP + 5 gms of vitavax 75% WP per tree by mixing it with 7-8 litres of water slowly along the tree trunk during May and repeat the treatment during September. Apply light irrigation after the treatment. To young plants the dose of fungicides and water may be halved, whole of the orchard may be treated once in a year.

(*iii*) Avoid flooding while applying irrigations. Modified basin system for young plants and segments for 4-5 trees for fully grown trees should be prepared for irrigation purposes. Do not grow Barseen or Potatoes in non-bearing pear plants.

(*iv*) While removing suckers from the underneath of trees avoid deep injury to the main roots. Control the growth of root suckers by spraying 500 ppm of Naphthalene Acetic Acid (NAA) during June-July.

PEACH

Botanical name

Prunus persica Batsch.

Family

Rosaceae.

Peach is an important stone fruit crop of the temperate zone. High quality peaches are produced in higher hills *e.g.* J.H. Hale, Alberta and Matchless peaches. In the plains during nineteen sixties only inferior quality peach 'Chakali' was grown. With the introduction of low chilling requiring peaches from Florida it has become the most important fruit of plains. These exotic cultivars ripen in May and June a month earlier than the indigenous cultivars, hence fetch good price in the market. These cultivars are being grown with assured irrigation facilities from Ganga Nagar (Rajasthan) to Dehradun (Uttarakhand).

Origin

It was thought that peach is native to Persia. But according to DeCandole, peach was in cultivation in China in 2000 B.C. It has been a common fruit of Mexico and England since long. Temperate zone countries have been cultivating peaches for centuries.

Uses

Peach is a delicacy as a fresh fruit. Peaches produced in temperate areas are more juicy and sweet than those produced in plains. New introductions for plains are sweet and acidic. Peach is a rich source of minerals and proteins. TSS of exotic peaches is less than Khurmani and Sharbati cultivars of plains. It can be processed as canned fruits, jam, nectar, juice and marmalades can be prepared with ease. It is laxative in nature just like pear and mango.

Climate and Soil

Peach prefers temperate climate. Low chilling requiring (250-300 hrs) peach cultivars prefer low temperature in winters and high temperature (40°C) in summers for good colour development and ripening.

Peach is not very fastidious to its soil requirements. Peach orchards are performing well from Abohar to Amritsar. The soil vary from sandy loam to clay loam, with good drainage. High pH (Kalar lands) is not suited to peach growth. It can grow well in soils with an electrical conductivity below 0.5 mm hos/cm, calcium carbonate less than 5 percent and lime less than 10%. Waterlogged soils are not good for peach cultivation.

Botany

The tree size varies from soil to soil and prevailing climatic conditions. In temperate zone, the tree size is medium but in plains the trees grows upto 6 meters of height and spread. The leaves are lanceolate or oblong lanceolate in shape. The leaf margins are finely serrated. Leaf lamina is shining on the upper side and light green from the lower side. Flowers are solitary, pink in colour, appear before leaves emerge. The sepals have pubescence. Fruits become soft at ripening. Epicarp pubescent. Stone hard may be clinging to flesh or remain free. These forms of fruits are called cling stones and free stones respectively.

There is another cultivated form of peaches, which produces fruits without pubescence or fuzz called nectarines.

Flowering and Fruiting

Peach flowers are insect pollinated. Honey bees improve fruit set to a considerable extent. Flowering starts in the first week of February and continue to end February. Peach cultivars being grown in plains are self-fruitful. The fruit follows a double sigmoid curve of growth just like pears. Peach fruit is a drupe like mango. Edible portion is epicarp and mesocarp. Endocarp is hard, stonoy which contain seed.

RECOMMENDED CULTIVARS

A. Yellow Fleshed (Exotic Cultivars)

1. Partap

Its fruit matures in the third week of April and takes about 75 days for maturity. This is the earliest ripening cultivar. Tree is of medium vigour. Average fruit weight is 70 gm. Fruit shape is almost round with red colour on the epicarp. Flesh is firm, free stone and acidic. Average yield per tree 70 kg. Juice contains 12% TSS and 0.7% acidity.

2. Flordasun

The tree is vigorous than Partap and is very precocious in bearing. It comes into bearing just after two years of planting the field. Its fruits ripen after Partap and fetches good price in the market. It is a free stone cultivar. It flowers profusely and fruit set is heavy, hence need fruit thinning for improving the fruit size. If the fruit is not thinned then the average fruit weight is 80 gram. But if fruit thinning is done it improves to 120 grams. It cannot be transported to distant markets due to its softness. It develops excellent reddish colour on the epicarp. Average yield per tree is 75 kg. The TSS of juice 12% with 0.47% acidity.

3. Flordaprince

Trees are as vigorous as Flordasun. Fruits have reddish colour on the epicarp on maturity. It ripens in end April. Flesh is firm and free stone at full ripe stage. It is prone to fruit fly attack. Average fruit yield per tree 100 kg. The TSS of juice 12% and acidity 0.5%.

4. Shan-e-Punjab

Tree in vigorous. The shoots show barreness of buds in the centre of growth. It flowers in the first week of February and fruits ripen in first week of May. Fruits are of bigger in size than all the peach cultivars. Average fruit weight 130 grams. Epicarp turns yellow with red blush on the tip. This cultivar is very much liked by the contractors and retailers because of its fruit firmness. Average yield per tree 70 kg. The flesh of the fruit is firm and become free stone at ripening. The TSS of juice 13% and acidity 0.56%.

5. Early Grande

Tree is semivigorous, high yielding and early maturing. Its flowering and fruit ripening coinsides with Shan-e-Punjab. Fruit size just at par with Shan-e-Punjab. Flesh of fruit is firm and free stone. Keeping quality is good TSS of juice 10.8 percent with 0.7% acidity.

B. White Flashed (Indigenous Cultivars)

1. Khurmani

Tree is of medium vigour, with upright growth habit. The foliage is usually dark green in colour than exotic cultivars. The fruits mature in second week of June. The fruits are medium is size with creamy epicarp. Average fruit weight 70 gm. Flesh is soft, juicy and cling stone. TSS of juice 13-14% with 0.4% acidity. Average yield per tree 70 kg.

2. Sharbati

Tree is vigorous. Flowering takes place in end of January. Fruits ripen end of June or first week of July. At ripening pink colour develops in the epicarp. It is a heavy yielder. Average fruit weight 70 gm. Average yield per tree 100-120 kg. The TSS of juice 13% and acidity 0.3%.

Flordaguard

This peach rootstock is resistant to root-knot nematodes.

Plant Propagation

Peach is commonly propagated by ring budding on peach seedlings. For raising of rootstock seedlings, seeds of desi (bitter) peach, khurmani or sharbati are collected. Seed has high seed germination rate and good seedling growth. Ring or T-budding is usually done in April-May more thick rootstock is usually tongue or cleft grafted during December-January.

Raising of Rootstock

Fruits of desi peach, khurmani or sharbati are collected in July. Seeds is extracted immediately and washed. Fresh peach seeds do not germinate, hence seeds need stratifications. Place the seed in layers on sand which should be kept moist from November to January. When the upper layer of seeds show germination, the emerging seedling are gently picked up and sown in the nursery beds. For ease sand is placed near a water tap and peach seed is mixed is the sand. The flowing tap water provide necessary moisture for the softening of the hard seed coat. Seeds start germinating after about 100 days in end January.

Stratification of seed can also be done by keeping the seed in cold storage at 10°C or at lower temperature for a month before it is sown in the nursery.

The planting of seedlings at two to four leaf stage during February is a sure method to get high success of seedlings in the nursery. Over 80 percent seeds germinate in the moist sand within 90-100 days of keeping the seed in open during November. This can be done by mixing seed with sand in October.

Care of Seedlings and Grafts

After transplanting the newly emerging seedlings in the nursery, light irrigation is applied. Seedlings are protected from the attack of white arts and aphids. The pencil thickness rootstock seedlings are budded in April-May. The rest of the rootstock is grafted in January. After budding/grafting of seedlings care should be taken to check bushy growth of buds in grafted plants. Allow only single shoot to grow as plant. To get sufficient growth of grafts the nursery may be provided sufficient N fertilization and frequent irrigation.

Crown Galls

Peach nurseries are infected with the formation of crown galls. This is caused by *Agrobacterium tumefaciens*. Tumours of varying sizes are seen at crown portions on roots or at graft joints. Follow the following nursery practices for the efficient control of the disease.

1. Procure fruits for seed extraction from a reliable resource, preferably from your own rootstock mother plants.
2. Avoid injury to roots and crown portion while uprooting and transplanting peach seedlings and doing hoeing etc.

3. Control effectively pests attacking nursery plants.
4. Uproot and destroy the diseased plants.
5. Discourage purchase of rootstock seedlings from nurseries showing attack of the disease.
6. While planting the seedlings, the diseased roots are pruned. Dip the roots in 5% of bleaching powder, before transplanting.

Planting

Peach plant go dormant in winter. It can be planted during dormancy, bare rooted. Healthy one year old peach plants are uprooted from the nursery along with full root system during December or first week of January. The nematode infected plants should be discarded. The formation of small root nodules on roots and sticking of soil onto the roots shows the attack of nematodes. Peach has short productive life span in comparison to pear and mango, hence should be planted as a filler in pear, litchi or mango plantations while planting keep the graft union at least 10 cm above the soil level. The height of the nursery plants should not be less than one meter.

Training and Pruning

To get high yield of good quality fruits the plants should be properly trained and pruned.

Training

Peach should be trained on modified leader system of training. However, it is difficult to find 10% plants in as orchard trained on this system, due to lack of knowledge. Most of the plants are trained on open centre system, thus poor yield at the full maturity of tree. Follow the following steps to train the peach plants on modified leader system of training.

1. Head back the plants to 90 cm at the time of planting. Cut back the apices of the side shoots. Remove all shoots up to 45 cm above ground level.

2. In spring and summers, many buds on the main trunk shall sprout, continue to check sprouts up to 45 cm of height on the main trunk. Select 4-5 side growing shoots by removing rest of the sprouted buds, by visiting the plants regularly.
3. In first January, *i.e.* after one year of planting. Remove other unwanted shoots by keeping already selected 4-5 scaffolds. Select 2-3 side shoots on the leader (top most shoot) and head it back. Now there shall be 6-8 scaffolds per tree. In summers head back these scaffolds to develop side branches. This will give the tree a shape of umbrella. The training of peach trees should be completed within two years.

Pruning

Peach bears on one year old shoots and a little on the short spurs (barbs). Therefore to get good growth each year plants have to be given pruning every year in December. Remove all dead and broken wood. The practice of thinning out of branches should not be followed. Thin out only those branches which grow inwards or crisis cross. This will help in maintaining sufficient fruiting wood for regular bearing of the tree. Heading back of 1/2 or 1/3 of all branches should be carried out on each scaffold judiciously. While heading back remove whole of the sterile part of the shoot. Otherwise the sterile portions of the shoot give ugly look and shall dry up from top. One can learn this with practice and after analysing the effect of pruning on the emergence of new shoots in March-April. Treat the major cut ends with Bordeaux paste. Severity of pruning can be adjusted according to health of fruit trees. Upward growing shoots at the top of the tree can be removed.

Fruit Thinning

Peach is a prolific bearer. There can be 2-3 fruits at one place on most of the branches. If all the fruits are left to mature, fruit size and quality shall be adversely affected. Overbearing in one year can bring alternate bearing in peaches. To overcome this problem, thinning of fruits should be carried out. Keep fruit to fruit distance 8-10 cms. Keep about 25-30 leaves per fruit on a branch. Thinning is done manually before 15th of March. Start thinning of fruits from base towards apex of the branch. Thinning operation must be completed before pit hardening of fruits.

Manures and Fertilizers

Peaches need sufficient nutrition to get profuse new growth each year. Since peach is pruned every year and lot of vegetative wood is lost, to maintain good vegetative growth sufficient nutrition is needed. Peach perform better on soils with high organic matter content.

Orchards where green manuring is done in the early years and 'Senji' is left to grow and die of its own in the orchard, produce high quality of peaches with the application of little dose of inorganic fertilizers.

Following doses should be applied for maintaining the health of plants.

Age (Years)	*F.Y.M. (Kg.)*	*Urea (46% N) (gm)*	*Superphosphate (16% P_2O_5) (gm)*	*Muriate of Potash (60% K_2O) (gm)*
1	10	100	100	50
2	30	200	200	100
3	40	350	400	250
4 and above	50	500	500	400

For sandy loam soils with poor organic matter content increase the dose of each fertilizer by 25 percent.

Apply farm yard manure plus super phosphate and Muriate of potash in December after prunings are removed from the tree basins. Mix these manures/fertilizers properly in the tree basins under full canopy of the trees. Do not apply fertilizers too close to the trunks. Leave at least 20-30 cm place near the tree trunks without fertilization. During the first year of the plant growth, split the urea into three equal portions and apply in February, March and April. Later on urea may be split in to two parts. Apply first half in February and second half after fruit set in March. Second dose may be adjusted as per bearing of the tree. Poorly bearing trees may not be given second dose to check laggy growth of shoots.

Iron Deficiency

It is very common is peach orchards. Peach planted on light soils or in soils with little higher pH often exhibit iron deficiency during May-June or after rains. It can be corrected by spraying Ferrous sulphate of laboratory grade @ 2g/L of water half the quantity of $FeSO_4$ lime may be mixed to make it more effective. Two sprays one in April and second in June are sufficient to check the deficiency.

Cultivations

No intercrop should be grown in peaches if these are planted as fillers in major fruit crops. If the peach is planted as main tree then pulses should be grown as inter crops. Some cover crops like cowpeas can be grown. Hoeing of the basins is enough to weed out the basins in December-January and again in April. No cultivation should be done in vacant places. The mixing of fertilizers shall provide sufficient aeration and weed control for peaches.

Irrigation

Peach requires frequent irrigations for fast vegetative growth and fruit development. Irrigation provided in January after application of manures in December shall provide sufficient moisture up to fruit set. Apply light irrigation after fruit set in end February. March to June apply irrigation at week's interval in light soils and at 7-10 days interval in heavy soils. Frequency of irrigation is dependant on type of soil and source of irrigation. The exotic peach cultivars *viz.* Flordaprince, Shane-e-Punjab, etc. which are early maturing need more irrigation from 10th April to fruit harvest. These cultivars may be applied irrigation after 4-5 days is May-June. Khurmani and Sharbati cultivars may be applied irrigation after 7-10 days. If rains are there the irrigation may be delayed. Water should not stagnant in the tree basins. The young plants may be supplied irrigation through modified basin system. Flooding of the whole area should be avoided for the first 4-5 years of age of the trees. Interval of irrigation may be increased to 20 days or so after fruit harvest. No irrigation may be given during leaf fall/dormancy.

Full grown trees need flood irrigation, hence the trees may be segmented into small blocks of 4-5 trees each. This will help in saving of irrigation water, loss of nutrients through leaching and protection from root injury due to root suffocation.

Weed Control

Weeds infest the peach orchards through out the year due to the supply of moisture. Perennial weeds like Dubh grass, Baru grass, Khabal grass, Parthenium and Daryabooti are common weeds. Annuals like Pitpapra, Bathu, Karund, Maina, Senji etc. grow from October onward. Tree basins are kept weed free by hoeing in December and February.

Gramoxone is the safest weedicide to be sprayed. The annual weeds should be sprayed at flowering stage with Gramoxone @ 6 ml/L of water. The perennial weeds need repeated sprays.

Glyphosate 41 SL can be sprayed @ 8-10 ml/L when the weeds have attained good growth. Avoid spraying glyphosate in tree basins. This weedicide should only be sprayed once in a year. Parthenium (congress grass) can be controlled by manually uprooting during rainy season or by spraying 20% common salt (NaCl).

Maturity and Fruit Harvesting

Peach follows a double sigmoid curve for growth during fruit maturity. Early maturing cultivars ripen the fruits in end April to end May. In same way harvesting continues for over a month. Harvesting should be managed in such a way that fruit of a cultivar do not get over ripe. Fruit maturity and ripening is cultivar specific. Partap and Flordasun need to be harvested faster than Shan-e-Punjab, which can be kept on the tree for few days longer. The following table gives, ample indication of fruit maturity and harvest in different cultivars.

		Time of			
S.No.	*Cultivar*	*Flowering*	*Ripening*	*TSS %*	*Acidity %*
1.	Partap	First week-Feb.	End April	10	0.56
2.	Flordasum	–do–	–do–	12	0.5
3.	Florda Prince	–do–	–do–	10	0.5
4.	Shan-e-Punjab	–do–	Ist week May	13	0.56
5.	Early Amber	2nd week Feb.	2nd week May	10	0.74
6.	Florda gold	Ist week Feb.	End May	11	0.83
7.	Florda belle	End Jan.	–do–	12	0.80
8.	Match less	2nd week Feb.	Ist week June	13	0.35
9.	Khurmani	–do–	2nd week June	13	0.45
10.	Sharbati	End Jan.	End June	15	0.34

Peach fruit should be harvested when still firm but have developed representative colour of the epicarp of the cultivar. Since ripe fruits which can not be transported to distant market should be sent to local market. The fruit picking should be done in plastic buckets/baskets to protect the fruits from bruised. 10-15 kg of fruit is taken into a gunny sheet ('Palli') and given few dips in water by tilting this way and other way to clear the fruits of fuss and dirt. Fruits are placed in shade and packed in plastic crates (20 kg) or in mulberry baskets (7-8 kg) by grading as per size of the fruit. Fruits are transported early in the morning to the local market.

However, to market peaches in the distant market, these are graded and packed in wooden boxes or card board cartoons and sealed properly. The use of grass or rice trash is made to protect the fruits from injury. Paper cuttings can also be used.

Commercially accepted fruit grades for peach and packaging size are given in table.

Fruit Grade	*Fruit Size (cm)*	*Inner Size of Box (cm)*	*No. of Layers*	*No. of Fruit per layer*
Special	5.5 – 6.3	37 × 16.5 × 16.5	3	28 – 32
Grade - I	4.6 – 5.5	–do–	4	35 – 38
Grade - II	below 4.6	–do–	4	38 – 43

Generally 2 and 4 kg corrugated fibre board boxes (CFB) are used for peach packaging. These are preferred over wooden boxes. Once used boxes for grape packaging are economical to pack peaches. Harvested peach fruits are required to be marketed expeditiously.

PLANT PROTECTION MEASURES

A. Insect Pests

1. Peach Leaf Curl Aphid (*Brachycaudus helichrysi*)

It is a polyphagous pest. It attacks the tender leaves, branches, flowers pedicles and calyx. It sucks cell sap from these vegetative parts. The leaves usually curl up and become yellow. The flowers and young fruits drop. Black sooty mould develop on its excreta, which hinder the photosynthetic activity. The pest is active from February-May.

Management

Spray Rogor 30 EC (Dimethoate) @ 2 ml/litre of water just before opening of flowers, or at leaf bud-burst or spray Nuvacuron 36% SL (Monocrotophos) 1ml/litre of water. Repeat the spray only if aphids are seen at fruit set or later on.

Peach Black Aphid (*Petrochlorus persicae*)

It is bigger in size than the green aphid and is black in colour. It congregate on stem and shoots or at tree trunks just below scaffolds. It sucks cell sap from the bark of limbs from March-June. If pressed is gives out reddish fluid. Its attack can devitalize the trees.

Managements

Spray Nuvacuron 36% SL (Monocrotophos) @1 $\frac{1.5}{2}$ ml/L of water on the colonies on stem and scaffolds. Black aphid collect and rest on shady side of trunk when sun is high.

3. Chaffer and Other Defoliating Beetles (*Adoretus Spp*)

Adult beetles appear with the break of monsoon and feed on leaves during night and hide during the day. In case of severe attack, fruits are also scrapped near the apical end. Beetles lay egg in the soil, grubs feed on roots and other organic matter. Sometimes grubs feeding on peach roots cause death of the plant.

Management

Spray Sevin 50 WP (carbaryl) @ 2 g/litre as soon as cuttings on the leaves are noticed. Repeat the spray after a week or Spray Dursban 20 EC (chloropyriphos) @ 2 ml per litre of water. Spray the insecticide in the evenings.

4. Hairy Caterpillars (*Euproctis Spp.*)

Sometimes the pest attack peaches also. Eggs are laid on the ventral surface of leaves by the females in clusters. On hatching, the young hairy caterpillars feed gregariously on leaf lamina. Whole of the mesophyll is eaten up, only mid-rib is left behind. Pest can feed on the epicarp of the fruit too.

Management

Spray Thiodan 35 EC (Endosulphan) @ 2 ml/L of water or Dursban 20% EC (chloropyriphos) @ 2 ml per litre of water as soon as the caterpillars are seen.

5. Peach Fruit Fly (*Bactrocera dorsalis*)

Fruit fly punctures the fruits when these are still hard and developing. The punctured fruits are not marketable, since fruits rot at the harvest time. When pressed the fluid oozoes out from the punctured site of the fruit. The larvae are seen under the epicarp in the pulp. Attack starts from mid-April to mid-May on the early ripening cultivars and in May-June in late ripening cultivars.

Management

(*i*) Burry the infested fruits deep in pits.

(*ii*) Hoeing the orchard in summers can kill the pupae which are present is the upper few cms of soil.

(*iii*) Spray Thiodan 35 EC (Endosulphan) @ 2 ml/litre of water or Sevin 50 WP (carbaryl) @ 2g/litre of water in the second week of April and repeat after 15 days twice up to harvest. Three sprays are sufficient to check fruit fly attack.

6. White Ant

It attacks the roots and climb up the trunk and scaffolds. If makes tunnels by feeding on the bark. Outwardly earth lines are seen on the damaged bark. The attack is usually after the fruit harvest, or in February-March. If not checked the trees can dry up due to severe feeding on the bark.

Management

(*i*) Notice the gullies on the trunk by periodically visiting the plants. Remove the earth with a broom or brush.

(*ii*) Apply one to two litres of solution per tree depending on the age of the tree. Prepare chloropyriphos solution by mixing Dursban 20% EC @ 10 ml/litre of water. Pore this solution slowly on the tree trunks so that it can percolate in the root zone.

7. Plum Case Worm (*Cremastopsychae pendula*)

The pest usually remain unnoticed due to its cap on the body. It feeds on leaves, bark of tender twigs and branches. It nibbles on the lamina and make holes in it. It can feed on the epicarp of the fruits.

Management

Spray Dursban 20% EC (chloropyriphos) @ 3 to 4 ml/litre of water or Sevin 50 WDP (carbaryl) @ 2g/litre of water as soon as the attack is seen.

B. Diseases

1. Shot Hole (*Stigmina carpophylla*)

Dark-brown scattered lesions appear on leaves, which enlarge rapidly. Abscission of diseased leaves due to shot holes take place.

Control

Spray Captan, Ziram or Thiram @ 2g/litre of water at leaf fall or bud swelling stage.

2. Bacterial Canker and Gummosis (*Pseudomonas monaprumonum*)

Bacteria attack all parts of the tree. There is a profuse gumming on the attacked bark of limbs and trunk. The limbs may dry up above the attacked point.

Control

Clear the wounds before rains and apply Mashobra paint, which should be prepared by mixing 225 g linolin, 12 g stearic acid, 150 g marpholin and 25 g streptocycline in 5 litres of water. Repeat the treatment on the new lesions in the following months.

3. Brown Rot (*Sclerotinia fructicola and S. fructigena*)

The fungus causes wilting and killing of petals and young twigs. The leaves wither and drop. Canker develops on large limbs, the bark get killed. Cracks open and gum oozes out. On fruits also gum oozes out and light brown spots appear as small round areas, which enlarge within a few days and entire fruit rots, shrivels and become hard, dry and hang on the trees as wrinkled mummy or fall on the ground.

Control

(*i*) Hanging or fallen fruits mummies should be collected and destroyed by burning or burying them in soil. The infected twigs and cankers should be cut out and burnt.

(*ii*) Avoid injury during picking. Select healthy unbruised fruits for the market.

(*iii*) Spray captan @ 2g/L of water 20 days before fruit harvest.

PLUM (PRUNUS SALICINA) FAMILY ROSACEAE

Plum is temperate in nature and thus a major crop of hills, but in north Indian plains and particularly in Punjab it is a minor fruit crop of great importance. It is preferred for planting as a filler fruit plant in mango, litchi and pear orchards due to low chilling requirement and less juvenile period. A few plantations of plum as solid blocks also exist in Amritsar, Fazilka. Ambala and Kurukshetra areas. Trees are highly productive due to profuse flowering, high fruit set and early ripening habit of cultivars recommended for plains. On an average a full grown tree can yield 50-60 kg of fruit. Plums are rich in sugars and vitamin-A. Fruits are eaten as fresh or processed into jam and squash.

Origin

Plum is native to China, although it became a commercial fruit of Japan and America. It is known as Japanese plum due to its cultivation in Japan from where cultivars spread to other places. At present, plum is cultivated in all temperate climate countries of the world.

Area

In north India high quality plum is being grown in hilly area of Himachal Pradesh, Jammu and Kashmir, Uttrakhand and Uttar Pradesh. In the plains low chilling requiring plum cultivars are cultivated through out Punjab, Haryana and in some parts of Uttar Pradesh and Rajasthan.

Uses

Plum fruit is rich in Vitamin A, B_1 (Thiamine), riboflavin and some minerals like calcium, phosphorus and iron. The well blended acidity with sugars is helpful in the preparation of jams and squashes. The dried plums are known as prunes. All plum cultivar can not be used for drying, that is why it said "all plums are not prunes but all prunes are plums." The prunes have great Ayurvedic medicinal value. Water prepared from the prunes is helpful in curing jaundice and summer bite.

Botany

Prunus salicina belongs to family Rosaceae to which peach and pear belong. Trees are of medium in size in the plains but in hills these are of small size. The bark of twigs and trunk is of blackish-brown in colour, usually shredding from the old scaffolds. Leaves are oblong obovate and sharply pointed. Leaf margins are serrated, lamina green above and light green on the lower surface. It flowers in first fortnight of February. The flowers are bronze on small spurs with very close rings. Thus flowers occur closely and inflorescence is termed as cymose. Fruit is a drupe (stone) with thin edible exocarp and fleshy mesocarp. The hard stony endocarp contains seed.

Climate

Plum prefers temperate climate. However, it has been found growing from higher hills in Srinagar to Jaipur in Rajasthan and areas around Delhi. It requires less chilling hours that is temperature below 7.2°C. It can tolerate frost and high summers both, that is why it can be cultivated in both low temperature to 0°C and up, highest up to 47°C in summers. Some cultivars like Santa Rosa can only be cultivated in higher hills (700-1000 chilling hours) but low chilling requiring cultivators (250-300 chilling hours) are preferred for plains.

Soil

Plum is performing very well in soils with high pH where peach fails. That is why it is preferred as filler over peach. For good performance of trees well drained sandy loam to medium loam soils are most suited. Root stock plays a major role in the preference of soil. Own rooted plum cuttings and Kabul Green Gage plum rootstock prefer heavy soils, where as plum propagated on peach do well in sandy loams. Plum propagated on peach should not be planted as filler in Litchi orchards, because of high water requirements of litchi plants.

Cultivars

There are many plum cultivars being cultivated in plum growing areas like Santa Rosa and Mariposa in higher elevations. For plains cultivars are discussed below :

1. Satluj Purple

It is a self-unfruitful cultivar and requires Kala Amritsari as pollinizer. Kala Amritsari plum should be planted as alternate plant in alternate rows for achieving good fruit set. Tree is medium in vigour with upright habit of growth. The fruit is of medium in size weighing 25-30 g. The epicarp develop crimson colour on ripening. The epicarp is thick and mesocarp yellow coloured firm flesh. The TSS of Juice ranges between 12-14 percent and acidity 0.5 to 0.7 percent. The fruits are usually eaten fresh. It is an early ripening low yielding cultivar. It ripens in early May and yield 35-40 Kg of fruit per tree.

2. Kala Amritsari

It is self-fruitful high yielding local cultivar. The trees are vigorous with profuse branching. This is most preferred cultivar of plains. Fruits are of medium size, round oblate depressed at both end. The epicarp (skin) on ripening turn dark purple. The mesocarp (flesh) is yellow in colour with juicy pulp. Fruits are little acidic in taste. The TSS of juice ranges between 13 to 15 percent and acidity 1.0-1.3 percent. Fruit ripen in second half of

May. Average yield about 55 kg per trees. Its fruits are preferred for making Jam and squash.

3. Titron

It is a self-fruitful cultivar but its yield increases if Alucha Early Round is used as a pollinizer. Titron trees are smaller than kala Amritsari. The fruit size is also smaller than both Satluj Purple and kala Amritsari. The epicarp is thinner than kala Amritsari. The flesh is yellow and less juicy. The taste is sub acid due to very low acidity. Fruit yield 30-35 Kg per tree. It is good for table purpose as well as jam making. TSS 13.5 percent and Acidity 0.4 percent.

4. Kataruchak

The cultivar originated in village Kataruchak of Gurdaspur district in Punjab. The cultivar is partially self-fruitful. The yield per tree increased where both kataruchak and Kala Amritsari had been planted. The fruit sell at a premium than those Kala Amritsari, due to the presence of white bloom (purbescence) on the epicarp (skin) of fruits. Trees are as vigorous as of Kala Amritsari. The fruit is bigger in size, heart shaped and purplish in color. It ripens just after kala Amritsari. Fruit yield 45-50 kg per tree. Fruit is good for table purpose, jam and squash.

5. Alubokhara

It is self-unfruitful and need 'How' as pollinizer. The tree is upright and spreading. The fruits are of larger size than all other cultivars. Its yield is less than kala Amritsari. The eqicarp yellow in colour with reddish spots. Pulp is juicy and sweet. TSS 15 percent, acidity 0.7 percent.

Propagation

1. Through Cuttings

Usually plums are propagated through of hard wood cuttings. Kala Amritsari is also propagated through hard wood cuttings. Cutting of Kabul Green Gage can also be used for grafting of other plum cultivars. The rooting percentage of cuttings can be increased by dipping the cuttings lower ends for 24 hours in 100 ppm of IBA solution before planting during December. Other cultivars can be grafted on to the cuttings of Kala Amritsari during December. Dissolve 100mg of IBA in 20ml of absolute alcohol and make the volume one litre. The success of these grafts ranges between 70-80 percent. Though grafts which fail to sprout can be regrafted on to the shoots developed from the rooted cutting next winter. The thickness of cutting should not be more than pencil thickness and length normally 18 to 20cm. Simultaneously grafted and rooted cuttings/grafts get ready for planting in the next winter.

2. Through Seed

Plum stones from the ripe Kala Amritsari fruits are extracted during May. These stones are properly washed to remove pulp and stored after room drying in gunny bags in shade. The seeds are sown in lines in the nursery during November. The seeds are covered with 4-5cm thick layer of sand. This will be sufficient to stratify the seed in field and seed will start germinating. Alternately the seeds can be stratified in wooden boxes as in peach and kainth seeds. Stratification is complete when the plumule emerges out from the stones of the upper layer of stones in the box. The emerging seedlings from the stones are gently collected and sown in lines 30cm apart at a distance of 10cm. Leave a gap of 45cm after transplanting four rows : Thereafter light irrigation through flooding should be given. These seedlings get ready for grafting within a year.

Planting

Plum plants in the nursery shed their leaves in January. Hence, these are uprooted along with the roots and can be transported long distances bare rooted. While planting the plants in the prepared pits, the lowermost branches should be removed. Keep only 3-4 side branches and clip the tops of all shoots so that secondary branching could be initiated. Place in the small pit all the roots and put soil to cover the roots to the soil level. Press the soil around the newly planted plants gentally. Apply light irrigation to the plants so that soil settle down. Level the soil at 'wattar' condition. Continue to apply light and frequent irrigation till rainy season.

Apply 10ml of chloropyriphos in one litre of water to each plant in the root zone 20 days after planting for the control of white ants. The plants getting tilted with first irrigation should be straightened and root zone repressed. Normally plums are planted as fillers, hence planting distance shall be according to the main fruit crop. This may vary from 3.5 to 5.0 m. However, if pure plum orchard is to be planted, then a spacing of 6m X 6m is enough, which will accommodate 256 plants in square system and 305 in hexagonal system of planting per hectare.

Training and Pruning

Plum trees should be trained according to modified leader system. If healthy plants with some scafforlds selected at the time of planting are planted, the training can be completed in two years. Otherwise, the training is completed in 3 years from planting. The top growing shoot is headed back at plating time at a height of 90cm. The scaffolds selected at a height of 40cm and above are also headed back in January to get branching. The plants sprout in February and continue to produce sufficient growth during summer until October. One should continue to select side branches during the growing period which will provide a healthy scaffold system within two years. In all, select 4-5 scaffolds all around the trunk. During the first two years selection of secondary branches on the scaffolds is also carried out. In the third year the top growing leader should be removed from a point of out growing scaffold, during January. Thin out some of the side branches and cris-crossing shoots. The small spurs formed during these years on shoots start flowering. Water shoots sprouting on the main trunk should be removed immediately.

Pruning

Plums bear on one year old shoots and short spurs depending upon the nature of cultivar. Normally no pruning may be given in the first year of bearing. Remove only cris-crossing shoots and water shoots. Do not too open the tree from inside in the first year of bearing. Remove only cris-crossing shoots and water shoots. Do not too open the tree from in side, let the thin branches have leaves which contribute toward the maturity of other shoots and spurs. A little pruning to the out growing branches may be done in fifth year onward. The severity of pruning depends upon the planting distance. When plum is planted as filler in pear it may require early severe pruning due to lesser planting distance than in litchi and mango. The cuts may be treated with Bordeaux paste/paint.

Flowering and Fruiting

Flowering in plum cultivars usually take place in the second half of February in the north India plains. Plum flower profusely hence there are sufficient number of hermaphrodite flowers to set fruit. Usually there is very heavy fruit set in plums. However, it is cultivar dependent. For example Kala Amritsar is heavy bearer but Alubokhara shy bearer.

Plum follows a double sigmoid curve of fruit growth. Immature fruits are very acidic. As the fruit mature the acidity start decreasing and TSS goes on increasing. Simultaneously the skin colour specific of cultivar start developing.

Irrigation

Plums are shallow rooted and fast growing plants hence need sufficient moisture during growing period. Interval of irrigation may depend upon many factors such as soil type, climate and kind of main fruit trees. Frequent irrigation at weekly interval may be applied during April, May and June. No irrigation at full bloom stage and the ripening stage be given to avoid flower and fruit drop. During rainy season no irrigation is required. The interval may increase to 20 days in September, October and November. There should be no irrigation given during December and January months.

Intercropping

In solid block of plums inter crops can be grown for the

first five years. If the plum is planted as filler then there is little space left for growing of crops. Hence innercrops can only be grown for 1-2 years. The legumes like peas, grams, moong or vegetables should be grown with caution. Water needs of intercrops should be such that also favour plum growth and fruiting.

Manures and Fertilizers

Since plum trees grows fast they require nutrition to maintain growth. Plum should be given following doses as per age.

		Dose per tree (g)		
Age (Years)	*Farm Yard Manure (kg)*	*Urea (N 46%)*	*Superphosphate (16% P_2O_5)*	*Muriate of Potash (60% K_2O)*
1 – 2	5 – 10	50 – 100	60 – 120	25 – 50
3 – 4	15 – 20	150 – 200	180 – 240	75 – 100
5 – 6	25 – 30	250 – 300	300 – 400	125 – 150
7 and above	40	300	400	200

Apply farm yard manure, super phosphate and muriate of potash in December. Nitrogenous fertilizer (urea) should be split in to two halves. Apply one half in February and second half in April. In addition the plum trees may be sprayed with a mixture of Zinc Sulphate plus ferrous Sulphate. Both should be of chemical grade. Spray 1½ kg zinc Sulphate plus 1½kg ferrous sulphate plus 1½kg unslaked lime in 500 liters of water during March just when fruits have become pea sized.

Weed Control

In plum orchards several weeds come up. Weeds particularly motha, Khabal grass, baru grass, parthenium and dab are found in abundance. Weeds may be removed from plant basins by hoeing. Weeds from the vacant space may be controlled by spraying Round up (Glyphosate) @ 10 ml/litre and when required particularly during the rainy season.

Thinning of Fruits

Generally plums set fruit much more than it is capable of

maturing. Excessive fruit set in plums results in to smaller fruits, which fetch poor price in the market. Kala Amritsari needs thinning of fruits to improve in size. Plum should be thinned immediately after natural fruit drop in April. Hand thinning is usually done from top of the shoot to bottom. Fruits are kept at 5-6 cm apart there should be 15-20 leaves for a fruit. There should not be two fruits at one point.

Fruit Harvesting

The plum fruits are mature when these have attained proper size and developed proper colour depending upon the cultivar. The plum fruits are harvested from second week of May in Punjab. For local marketing fruits should be harvested when ripe and firm. For distant markets, fruits are picked when firm but have developed 50% colour on the skin. Plum should be harvested along with pedicels avoiding any injury to the fruit. Plum fruit is very perishable in nature, hence should be handled with care. The small baskets should be padded with rice trash or grass at the bottom and sides. Freshly harvested fruits are transferred in these baskets and covered with paper and tied in guny cloth. The fruits should be graded before packing in basket or wooden boxes. Several pickings are made as the entire fruit on a tree do not ripen at one time. Fruit is borne on spurs also so care should be taken to the save the spurs from breakage, during harvesting.

Post-harvest Handling

Different grades of fruits are packed in different boxes and labeled. Three standard sizes as per "Ag-Mark" System of grading plum are as follows :

Grade	*Fruit Size (diameter cm)*	*Inner Box Size (cm)*	*Number of Layers*	*Number of Fruits Per Layer*
Special	4.2 and above	36 × 16 × 16	3	28 – 32
Grade-I	3.6 – 4.2	Do	4	38 – 42
Grade-II	Below 3.6	Do	4	50 – 56

For distant markets wooden boxes are preferred over basket. To save fruits from injury each layer of fruit is covered with paper strips and newspaper sheet. Finally the lid of the box is nailed.

PLANT PROTECTION

A. Insect-Pests

1. Aphid (*Brachycaudus* helichrysi)

It attacks the young emerging leaves and shoots. It sucks the cell sap from the young tissues. Severe infestation affects the growth of shoots and fruit set.

Management

Spray the plants at bud burst with Rogor 30 EC (Dimethoate) @ 2 ml/L or Nuvacuron (*Monocrotophos*) 1.5m/L and repeat after fruit set.

2. Plum Case Worm (*Cremastopsychae* pendula)

The name shows the pest has a covering of conical shape. It nibbles on the tender bark, leaves and even fruits.

Management

Spary 1kg of Sevin 50 WP (Carbaryl) or 750 ml Thiodan in 500 litre of Water.

3. Hairy Caterpillars (*Euproctis* SP)

These are sporadic pests. Females lay eggs in clusters on ventral surface of leaves and covered with yellow hair. On hatching the young caterpillars feed gregariously on leaf lamina, skeletonizing it completely. Later the caterpillars segregate and gnaw the leaves. In case of severe infestation, the entire tree may be defoliated. Young caterpillars have whitish hair, while full grown cater pillars have red head, dark-brown body with white hair on the head and a tuft of long hair at anal end. Adults are yellow moths with pale transverse lines on forewings.

Management

1. Collect and destroy the egg masses and gregariously feeding young caterpillars.
2. Spray 750ml thiodan 35EC (endosulphan) in 500 litres of water.

4. Plum Defoliating Bettle (*Anomala lineatepennis*)

Bettles are polyphagous in nature. The bettle attacked leaves presents a characteristic sieve like appearance on account of shot holes. Beetles are nocturnal in habit and cause damage at night time.

Management

1. Spray sevin @2g/L or nuvacuron 36SL (monocrotophos) @ 1.5ml/L against adult beetles.
2. Apply phorate or aldicarb 2kg/hectare.

B. Diseases

1. Shot Hole (*Stigmina carpophylla*)

It is caused by a fungus stigmina carpophylla, symptoms appear both on leaves and twigs. Dark-brown lesions appear on leaves and enlarge. The brown area fall giving a form of shot holes in leaf lamina. On twigs and fruits the lesions develop into canker. The affected twigs dry up, the fruit may fall.

Control

I. Remove the dried branches. Collect the affected fruits and destroy.

II. Spray Captan on Thiram @ 3g/ litre at leaf fall stage and repeat at bud swell stage. Continue the spray in the next year too.

2. Bacterial canker and Gummosis (*Pseudomonas monaprumonum*)

The bacteria attack every part of the tree including tree

trunks, branches, shoots, spurs, blossom, bud and leaves. The attacked limbs get girdled when the infection is severe. The limbs above the attacked point dry up.

Control

Clean the wounds and apply mashobra Paste before the start of rains. For 5 kg of Mashobra paste dissolve, lanolin 225 g, stearic acid 12 g, marpholin 150 g, streptocycline 25 g in 5 litres of water.

GUAVA

Botanical Name

Psidium guajava

Family

Myrtaceae.

Guava is an important fruit of tropical and subtropical area of the world. It is commonly called poor man's fruit. Guava contains maximum vitamin C content per 100g. of pulp after amla. It contain antioxidant factors and can control systolic blood pressure. It is good source of roughage and help in removal of constipation. In India it is cultivated almost through out except higher hills. It gives two crops in a year. That is why it is being preferred for plating in soils where citrus can not be grown. Two main cultivars which originated in Allahabad and Lucknow in Uttar Pradesh occupies most of the area under guava. Guava has become an important fruit crop of Punjab and its area ranks second after citrus. It grows well throughout the states of Punjab, Haryana and Uttar Pradesh.

Origin

Guava is native to Central America and is widely cultivated in South Africa, Hawai, Indian continent and Maxico.

Area

In India guava occupies an area of 2.03 lac hectares with annual production of 22.7 lacs MT. Allahabad area in U.P is reputed for the production of high quality of guava in India and the world. In Punjab it occupies an area of 8022 hectares with annual production of 160463 MT.

Uses

It is a big source of Vitamins C, A, B_2 (riboflavin) and minerals like calcium, phosphorus and iron. It contain about 180-300 mg of vit. per 100 g of pulb. Ripe guava fruits contain 14 percent TSS; 0.3 percent acidity and 7 percent fibre. Guava fruits are used for Jam and Jelly making. Its juice is used to blend the pear and peach juice. Guava fruits are used as mixed fruit chats in parties.

Botany

Common Guava (*Psidium guajava*) is cultivated for its edible fruits. Edible portion is thalamus and pericarp. Some other species of genus psidium are ornamental cattley guava (*Psidium cattleianum* Sabine), a shrub Brazilan guava (*Psidium guineese* SW), acidic Chinese guava [(*Psidium friedrichsthalianum* Berg.) Nied] and mountain guava (*Psidium montanum* Sw).

Psidium Guajava

It is a spreading tree, may grow up to height of 8 metres with same spread. The trunk is light brown in colour with shredding scaly bark. Highly branched tree. The leaves are oblong-elliptic in shape. The pubescent is present on the underside of leaves. Leaves are light green to dark green in color show crimson coloured strips during cold weather. Flowers appear on new growth singly or in three to four. Ovary is inferior, stamens many. Fertilization takes place before the opening of the flower, which is termed as *cleistogamy*. The fruit shape may vary from round to oblong or pyriform depending upon the cultivar. Flesh is soft contains many seeds in the centre, although seedless fruits are also found.

Climate

Just like mango, guava is found growing from plains to higher hills owing to its wider adaptability. It likes mild rainy season during fruit maturity and ripening just like mango. More rains in a year may lead to profuse growth with decreased productivity. It can tolerate mild frost in winter and drought in summer. It requires dry climate condition at the time of flowering.

Soil

Guava is considered to be hardy fruit crop. It is not very choosy for soil. It can grow well in slightly alkaline and poor soils. For good growth of trees well drained sandy loams to clay loams with 6.5 to 8.5 pH are best suited.

Cultivars

Some important guava cultivars are discussed below :

1. Allahabad Safeda

The trees are vigorous, compact, taller than Lucknow-49, usually with round crown and dense foliage. Grows upto 7-8 metre high. It has the tendency to produce long branches. The fruit vary in size depending upon the fruit load on a tree. The fruit shape usually round, some time with short neck pyriform (Pearshaped strain), smooth surface. There are pink dots scattered over the whole fruit, (Chittidar). Pulp is soft and melting, white, taste sweet. Quality of fruit is excellent. It takes about 75 days to maturity. TSS 13 percent. Full grown trees produce over two quintals of fruit annually (Two to three crops).

2. Lucknow-49 (Sardar Guava)

It is a selection from Allahabad Safeda. The trees are spreading with drooping branches. It is dwarfer than Allahabad safeda have dense foliage. The trees grow at maturity upto the height of 6 metre. Fruit size depends upon the load of the crop. Normally it is round in shape with little rough surface. Fruits light green to creamy at ripening. The number of seeds is more than safeda. It is heavy bearer. Yield per tree annually, may be 2 quintal or more. TSS of juice 12.5 percent.

3. Red Fleshed

It also originated in Allahabad area. The tree is vigorous and spreading mature tree can have 7.0 metre of height. The crown is uneven due to long breaches and dense leaves. The trees bear flower through out the year, hence fruits are available through out. The fruit flesh is pink in colour with many seeds. Fruits are smaller in size than Lucknow and Allahabad safeda. TSS of juice 12 percent.

4. Seedless

The tree grows upright with sparse foliage on the shoots. Leaves are usually larger in size than other cultivars. Fruits are pyriform in shape. Skin colour yellow at ripening. The epicarp may show ribs. Pulp is white with few seeds. The yield is low, 70 kg per tree annually. TSS of juice 12 percent.

Other new promising guava varieties shewta, Lalit, Punjab pink, Arka Amulya, Pant Prabhat.

Raising of Rootstock

Propagation of uniform material for planting is very important. For the vegetative propagation of guava, rootstock is raised from the seeds extracted from Lucknow-49 (Sardar guava). The seed is extracted from ripe fruits. The seed is sown in raised beds closely immediately after extraction in August or in February. Seedlings of 10-15 cm height are uprooted from these beds and replanted in well prepared field for patch budding, which is done on pencil thick root stock seedlings.

Patch Budding

Angular budwood of current season growth is used. The scion buds of IX2.5cm is removed carefully and fixed on exposed portion of root stock of same size and tied with polythene strip leaving both the buds uncovered. Both buds can sprout. Choose one of the shoots to develop into the plant. The most suitable time for patch budding is May-June.

Vegetative Propagation

Guava which is usually propagated from seed can also be propagated vegetatively. The plants can be raised through cuttings, layering or mound layering.

Air Layering

Air layering with the use of growth regulators has been reported to be the best method of guava propagation. For this purpose one year old shoots of uniform diameter (1cm) are selected. A ring of bark 2cm in length just below a bud is removed about 30-40 cm behind on the selected shoot. Apply IBA 500 ppm or NAA 500 ppm solution at the apical side of ringed bark. Cover the ringed portion as well 5cm further shoot with moist sphagnum moss and wrap it with 15 cm wide polythene sheet. Tie the both ends of the sheet of the air layered shoot. Air layering can be performed in March. Many roots are visible in the polythene sheet after 35-40 days of the operation. Remove the air layers from the mother plant in July and plant in the nursery by carefully removing the wrapped polythene. It will be better if ½ way cut is given to rooted layers 7 days before severing from the parent plant, for planting in the nursery.

Rooting of Cuttings

Guava can easily be propagated through cuttings. Cutting of 20-25 cm length are prepared from one year old shoots. Dip the cuttings for 24 hours in 200 ppm IBA solution before planting in the prepared beds. Cuttings root well during February-March. The planted cuttings should be applied light irrigation. Keep the soil medium at field capacity by spreading trash for the conservation of soil moisture.

Mound Layering

Guava is also propagated through mound layering, instead of air layering. For this purpose mother plants are planted at

1m X 1m apart. The plants are headed back during July-August by keeping a stub of 5-8 cm above the ground . A number of shoots emerge on the stub. A ring of bark just as in air layers is removed in March and a mound of moist soil is prepared to cover the ringed shoots. Roots emerge above the ringed portion with in 30-40 days. These stools are ready to be detached from the mother plant in July-August, from where again many shoots develop which are reused. Same plant can be used for many years. At planting time of layers 30 percent of heading back shall improve establishment success.

Planting

Guava is usually planted as compact blocks at a distance of 7 m apart and thus require 196 plants on square system and 216 on hexagonal system of planting. But it is also being planted as a filler in litchi and mango orchards, thus number of plants shall vary according to the planting distance of the main fruit crop. Guava being evergreen can be transplanted in both the seasons, *i.e*, rainy season August to October and spring February-March. Care should be taken while removing the earth balls, that tap root of the rootstock should be cut on a depth of at least 20-25 cm. In case of vegetatively propagated plants all roots should be excavated. Guava can also be planted bare rooted during September-October. In this case plants may be defoliated at the time of removal from the nursery and roots covered with moist trash (paddy trash) and packed with gunny cloth to check transpiration.

Training and Pruning

Guava should be trained on modified leader system of training, which can easily be completed in three years as in peaches. This will provide a strong frame work to the guava trees which will help to support high yield. Since, guava bears on current season's growth hence, a light pruning at the time of first harvesting will be beneficial for maintaining the productivity of the tree. Remove 10-15 cm of some of the shoots in a tree while harvesting the fruits. Broken and dead shoots should be removed regularly.

Manuring and Fertilization

Age of tree (Years)	*Farm Yard Manure (kg)*	*Urea (46% N) (g)*	*Superphosphate (16% P_2O_5) (g)*	*Muriate of Potash (60% K_2O) (g)*
1 – 3	10 – 20	40 – 120	100 – 300	40 – 120
4 – 6	25 – 40	160 – 200	400 – 600	150 – 225
7 – 10	40 – 50	250 – 300	700 – 1000	250 – 350
Above 10	50	300	1000	400

Farm yard manure should be applied in tree basins during May, only once in a year. Since, Guava bears three crops the flowering over laps each crops, hence the above recommended dose of inorganic fertilizers N and P should be applied to each crop in June for rainy season crop and in September for winter fruiting. Muriate of potash can be omitted in September.

Zinc and Iron Deficiency

The leaves show yellowing and chlorosis in between the veins. It will be beneficial if one or two foliage sprays are given. Spray the trees with 2 gm/L of water mixture of Zinc Sulphate, and Ferrous Sulphate in July and again in September. If this spray is regularly given the plants remain healthy and bear good crops.

Irrigation

The newly-planted guava plants should be given irrigation at an interval of a week during summer and 15-20 days during winter. The bearing trees should be given irrigation at interval of 10-15 days in summer and 15-20 days in winter depending upon the type of soil and its water holding capacity. The modified basin system of irrigation should be preferred over flood irrigation. No water should stagnate as it causes root rot. During frosty nights light irrigation should be applied.

Flowering and Fruiting

Flowering in guava continue throughout, starting from April May to December-January. The trees flower even when the green fruits are still there. Thus trees may bear three crops in a year, *i.e.*, rainy and winter and winter in North Indian conditions. The rainy season crop is usually heavy but get infected by fruit fly. Guava bears flowers solitary or in cymes of to 5 flowers in the axil of the leaf on the current growth. The blooming period varies from cultivar to cultivar and season of flowering. Due to cleistogamy the fruit setting is very high. Negligible flower drop has been observed. Young fruits drop due to competition, however 3-4 fruits can mature at one point. The rainy season crop is harvested in July-August and winter crops in November-December and then in March, Guava fruits take 120 to 140 days to attain maturity for harvest.

Intercropping

The vacant space between the plants should be judiciously used for growing of leguminous crops like cowpeas, Guara, gram and moong, etc. Wheat can also be grown but separate irrigation system should be there to apply irrigation to guava plants in March-April. Vegetables like carrot, turnip, radish and peas can also be grown.

Weed Control

Weeds should be kept under control through out the year because fruit harvesting is done through out the year, for the easy mobility of labour. For this purpose, spray Gramoxone 6ml/litre of water during March, July and September. To control perennial weeds the March spray may be done with Round up (Glyphosate) @ 10 ml/litre. Spray of these weedicides should be carried out on a calm day to avoid the damage to guava foliage.

Crop Regulation

Guava gives two to three crops in year. The growers have been advised to for go the rainy season crop as it is usually attacked by fruit fly. However, orchards are auctioned for the whole year hence it becomes difficult to convince the contractors to eliminate the rainy season crop. In order to get only winter season crop following method should be adopted with the connivance of the contractor :

1. Spray urea 10 percent or naphthalene acetic acid (NAA) @ 600 ppm at full-bloom (May). Dissolve 120 g of NAA in 300-400 ml of absolute alcohol, slowly pore this solution into Drum containing sufficient water make the volume to 200 litres. Prepare more solution as per the requirement of the area to be sprayed.
2. OR Prune 20-30 cm of the apical portion of shoots during last week of April. This will avoid rainy season crop completely.
3. Follow the recommended fertilization schedule and with hold irrigation during April-May.

If the orchard is in frost prone area, then one may not opt for only winter crop. It has been seen that total winter crop can be lost due to frost injury during December-January.

Harvesting

Guava fruit is climacteric in nature. It can be picked when mature, light green in colour and firm. It ripens during transport on reaching the market and gives excellent colour and taste. The contractors usually harvest the fruit when ripe or nearly ripe for direct sale to the consumers. Few leaves or small branches if kept with the fruits make the fruit attractive and trees get light pruning. For distant marketing this practice should be discouraged.

Post-harvest Handling

Ripe guava fruits are perishable in nature. Fully mature, light green and firm fruit are harvested and placed in shade in a verandah for grading and packing. For packing CFB cartoons of 4 to 10 kg size are preferred over the bamboo baskets of different size. Guava fruit can be kept at room temperature for a week in perforated polythene bags during winter and for three weeks in CFB cartoons in commercial cold storage at 5-10°C and RH of 85-90 percent.

PLANT PROTECTION

A. Insect-Pests

1. Fruit Fly (*Bacterocera dorsalis*)

Fruit flies are the most common and serious pest of guava. The flies lay eggs on the surface of fruits at colour break stage. On hatching the maggot enter into the fruit and feed on soft pulp. When fruits are cut open the white maggots are seen in the flesh. The infested fruits rot and fall down. *Pupation* take place in the soil.

Management

1. Clean cultivation/sanitation of orchard is important to reduce/check fruit infestation of fruit flies. Collect and dump deep in pit the fallen infested fruits. Daily cover the fruit with soil. Do not leave the pit uncovered for long.
2. Avoid taking rainy season crop if possible.
3. Plough and stir the basins during May-June to kill the pupae.
4. Harvest the fruits when light green in colour and firm.
5. Give two sprays of thiodan 35EC (endosulphan) 2ml/ per litre or Sevin 50WP (carbaryl) 2g/litre, one before maturity and second at maturity. No spray should be given at ripening stage during harvesting of fruit.

2. Bark Eating Caterpillar (*Indarbela tetraonis*)

Insect bores in crotches of scaffolds and feed inside the bark and wood by making holes. Severely infected branches may dry up.

Management

Inject Kerosene oil or 50:50 chlorophyriphos: Water solution inside the holes and close the holes by plastering with mud.

3. Guava Shoot Borer (*Microcolona technographa*)

It is a serious pest in the guava nursery. It damages the tender shoots. The infested shoots dry up which can be located from a distance by the presence of fine black growth on the leaves.

Management

Spray (monocrotophos) 36EC Nuvacuron @ 2.8 ml per litre or (chloropyriphos) 20EC Dursban @ 4.5 ml/litre or Quinalphas @ 4 ml/litre of water to manage the borer.

4. Mango Mealy Bugs

See under mango.

Diseases

1. Guava Wilt

Wilt in guava is caused by a fungus *Fusarium solani* or *Cephalosporium* sp. or *Rhizoctonia* sp. The symptoms appear on the infected trees many months after their roots have been attacked by the fungi. Sparse foliage, yellowing of leaves and tree wilting are the symptoms. Before wilting tree may flower profusely and set fruit which remain small. Overbearing is due to stress of damaged roots. In the roots the cambium between the bark and wood show discolouration. Replant trees in the same pits bear fruit for few years and are again attacked by the fungi. Refilling of the gaps should be done after treating the soil with fungicides.

Control

Plant guava in well drained fields. Avoid flooding the guava field while applying irrigation. Apply 15 g Bavistin to each plant trunk in 2 litre of water.

(*i*) Uproot and burn the wilted trees along with all roots.

(*ii*) Drench the soil in the pit with 2 percent formalin solution and cover with sarkanda and old wetted gunny bags. Expose the soil for 14 days and replant healthy guava plants.

2. Fruit Rot, Anthracnose or Die Back

This is caused by fungi *Gloeosporium psidii* and *Phytophthora parasitica*. Fully mature fruits are more prone to fungal attack. The centre of a lesion has a pink sticky spore-mass characteristics of the anthracnose disease. Fruits rot completely within 2-3 days. The fungi also attack the twigs and branches of the tree during rainy season, resulting in the die-back of shoot from the distal end.

Control

(*i*) Rain or irrigation water should not be allowed to stand in the tree basin around the trunk.

(*ii*) Prune the infected parts of shoots/branches and spray Bordeaux mixture 2:2:250 or Blitox 3g/litre of water repeat the spray after 20 days of the fruit set and continue till maturity of fruit.

(*iii*) The rotten and mummified fruits which fall on the ground or may remain attached on the tree should be removed and burried deep into soil.

GRAPE

Botanical Name

Vitis Crinifera L.

Family

Vitaceae

Grape is an important commercial fruit crop of south India. Grape growing has been regarded as most remunerative enterprise. In North India particularly, Punjab its cultivation is being taken in a big way. Due to heavy initial investment on erecting the system of training and occurrence of rains at the ripening time of grapes, the area under grapes may not increase further. In Madhurai area through staggered pruning fruit is taken throughout the year.

Origin

Vitis vinifera is the most important species grown in India. Grapes originated in Armenia near Caspian Sea. In India, grapes were introduced from Iran and Afghanistan by the end of 12th century. However, Aryans knew grape cultivation and its uses much earlier. S. Partap Singh Kairon Chief Minister of Punjab imported Anab-e-Shahi and Perlette cultivars during 1962 from California.

Area

Area under grape in India is 79.6 thousands hectares with an annual production of 1878.3 thousand M. tonnes. Most of the area of grape cultivation is in Maharashtra, Andhra Pradesh, Karnataka and Tamil Nadu. It is being cultivated in North Indian states as temperate crop. Punjab occupies an are of 777 hectares with annual production of 22088 MT.

Uses

Generally grapes are eaten fresh as table fruit. As it contains glucose as major sugars it is easily digestible. Grapes are good source of calcium, phosphorus, iron and vitamins. The total soluble solids in different cultivars may range from 10 to 20 percent and acidity 1 to 4 percent. Fresh grapes are laxative in nature. Wine can be prepared from grape juice. Berries are used to prepare rasin.

Botany

Vitis vinifera L. belong to family Vitaceae. Genus Vitis includes two Sub – genera *viz.* Muscadina and Euvitis vitis vinifera has characteristic of shredding bark, forked tendrils, lobed leaves, elongated clusters, seeded or seedless berries remaining attached to the pedicels. Whole berries (pericarp and placenta) are edible.

Climate

Grapes are grown in both temperate and tropical climate. In tropical climate vines remain evergreen. Grapes performs well where there are no or little rain at the ripening time of the grapes. Usually dry weather during flowering and ripening is preferred for high quality grape production. Regions having heavy rainfall or hail storms are not suited for grape cultivation.

Soil

Grapes are found growing on a variety of soil type. Best soil for grapes would be well drained loam to sandy loam with good organic matter. Poorly drained, alkaline soils should be discarded. Grapes do well on soils having less than pH 8.7 and EC upto 1.5 mmhos/cm, calcium carbonate up to 10 percent and lime concentration up to 20 percent.

Cultivars

Just like in mango, there are over one thousands grape cultivars. However, only few are under wide cultivation. Some of the cultivars are discussed below :

Perlette

Perlette is hybrid between *Scolokertek hiralynoje* and Sultanina Marble developed by H.P. Olmo while working at Davis, University of California in 1936. This is early ripening seedless cultivar which was released for cultivation in Punjab in 1964. At present this is the only cultivar under cultivation through out North India. It bears heavily on short spurs (3 to 4 buds). The bunches are usually compact, prominently shouldered, conical and big in size. There may be over 300 berries per branch. Berries are green to amber in colour, medium in size and round in shape. The ripening starts from the beginning of 2nd week of June from Sirsa Hissar belt to 3rd week of June in Himachal Pradesh and Jammu and Kashmir. The T.S.S. of juice ranges between 14 to 18 percent with 0.5 to 0.7 percent acidity. It is prolific bearer and can yield 35 tonnes per hectare in bower system of trained orchard. The cultivar has two drawbacks, *i.e.,* compactness of bunches and shot berries. These draw backs have been over come by thinning of berries and bunch dipping in 40 ppm GA_3 solution.

Thompson Seedless

This cultivar was also released for cultivation along with perlette. But due to berry shatter problem its cultivation was stopped. The bunches are long and conical. The berries become golden yellow in color at full ripe stage .These are elongated in shape and are seedless. It ripens in end June. The TSS of Juice ranges between 15-20 percent with 0.5 to 0.6 percent acidity in north Indian plains. The yield is about 20 tonnes per hectare. Now the area under this cultivar is again increasing in north India due to its better quality grapes and changed climate.

Anab-e-Shahi

This is a vigorous cultivar and do well only on bower system. The bunches are attractive and bigger than both the above cultivars, Berries are well sized and become amber in colour when ripe. Berries have bold seeds, which are not edible. It ripens in July under north Indian conditions, T.S.S. of juice varies from 13 to 16 percent with 0.5 to 0.6 percent acidity. Average yield on bower system in northern plains ranges between 35 to 40 tonnes per hectare. In Punjab most of the area has been uprooted due to its late ripening and berry damage due to rains.

Flame Seedless

It is performing very well in the plains of Punjab and Haryana. The vines are vigorous than perlette. Bunches are medium in size and well filled, with light purple, seedless berries. T.S.S. of juice varies between 16 to 18 percent with 0.6 to 0.7 percent of acidity It ripens in mid June. It is an excellent table grape cultivar. The average yield per hectare on bower system is around 25 tonnes. The cultivar can do well on head system of training also. For uniform colour of berries treat the bunches which 400 ppm ethephon at colour break stage.

Beauty Seedless

It is a semivigorous cultivar. It is an early ripening cultivar, which ripens in mid June. Bunches are of medium in size, with bold, seedless bluish black berries. Berries are responsive to GA3 treatment. The TSS of juice may reach to 18 percent.

Kishmish Charni

This is a seedless cultivar from Uzbekistan. Very good table purpose cultivar. It bears big bunches like Perlette but less compact. Berries are brick-red in colour and are variable in size; Pulp is pinkish yellow, very sweet with 20 percent TSS and little acidity 0.4 percent. It can yield over 20 tonnes per hectare.

Some new cultivars have been developed at Indian Institute of Horticulture Research, Bangalore; IARI New Delhi and Agricultural universities. The promising hybrids, *viz.*, Arkavati, Arka Kanchan, Arka hans and Arka Shyam have been released for cultivation.

Propagation

In India all the grape cultivars are propagated through cuttings. For this purpose, hard wood cuttings from one year old shoots are prepared. Only healthy cuttings should be planted for rooting purposes to get healthy vines for planting. Cuttings should be prepared from the pruning wood during January. The length of the cutting should be between 20-25 cm with at least three nodes (buds). The lower cut should be near the bud and upper cut away from the bud. The cuttings should be tied in bundles of 100 cuttings each and placed in moist sand for 15 to 20 days for callusing. These cuttings are then planted in the nursery in the first week of February by keeping 1/3 of cutting above ground.

The planted cuttings are applied light irrigation at regular intervals to keep the soil in 'watter' condition for most the time for high percentage of rooting. Weeds are kept under control by hoeing. To prevent attack of white ants treatment of chloro-pyriphos @ 2 litre/acre is given after 20 days of planting of cuttings through irrigation water and is repeated in April-May.

Planting of Vines

The rooted cuttings of grape vines are transplanted in the prepared pits in January, before sprouting. The pits are dug at 3mx3m apart the pits are re-filled with mixture of 50 kg of FYM + top soil of the pit. At the planting time of the vines, keep only two to three buds on the new shoot. This will encourage to produce vibrant shoots at sprouting of the vine. Treat the whole field with Dursban 20EC (chloro pyriphos) @ 2 litre per acre at the time of first irrigation to control white ants attack. The same, treatment may be repeated in August-September.

The spacing of vines differs greatly depending upon the cultivar and training system. The following table gives number of vines per hectare.

System	*Planting Distance*	*Number of Vines Per Hectare*
Head system	2m × 1.5m	3300
Kniffin system	3m × 3m	1089
Bower system	3m × 6m	561

Training and Pruning

Training method differs with the system of training, *i.e.,* bower, kniffin, telephone, y-system or head. For high returns train the vines on bower system. Take the vine straight upto the height of the bower in the first year itself. This is possible only if the trainer visit the field periodically. Regularly pinch off the sub/side shoots emerging on the growing vine. Pinch the top of vine just below the bower level to encourage emergence of laterals. Select two laterals at the top and allow them to grow on either side, in opposite direction along the central wire. These two laterals shall form the primary arms. The length of these arms and secondaries on these arms should not be extended too fast. Regularly head back the arms to develop sufficient secondaries. The entire area on the bower for the framework of the vine should be covered in installments. The whole training of the vine should be completed within two years of planting of vine. The training should be done by a technical person.

Pruning

Recommended pruning intensity for different cultivars in Punjab is as shown on next page.

Pruning should be done in January every year when the vines are still dormant. Keep only healthy fruiting canes with the recommended number of buds per cane depending upon the cultivar. The rest of shoots are pruned to one or two buds. These are called renewal spurs. These spurs sprout to give healthy shoots which are selected as fruiting canes in the coming years. A balance 1 : 2 is kept between fruiting cane and renewal spurs. In Madhurai area staggered pruning is done to get fruit through out the year.

Name of Cultivar	*Planting Distance (m)*	*Training System*	*Number of canes per/vine*	*Number of buds per cane*
Anabe-e-shahi	3 × 6	Bower	80 – 100	6
Flame-Seedless	3 × 3	Do	50 – 60	5
Perlette	3 × 3	Do	60 – 80	4
Beauty seedless	3 × 3	Do	60 – 80	4
Kishmish charni	3 × 3	Do	60 – 80	4
Thompson seedless	3 × 3	Do	40 – 50	8

Double Cropping in North India

Vines go dormant due to severe winter, in north India. To take double crop on some cultivars like Beauty seedless and Perlette, vines are given a second pruning during end June or first week of July immediately after the harvest of the summer season crop. The autumn crop is harvested in October. The foliage and bunches are protected from anthracnose attack by regularly spraying with 2 : 2 : 250 bordeaux mixture and Bavistin @ 1 gm/litre of water alternately.

Berry Thinning

Perlette cultivar have compact bunches. To loosen the bunches and reduce the shot berries in the bunch following practice may be adopted.

1. Thin the flower buds a week before full-bloom by leaving 100 to 120 flower buds per panicle.
2. When the berry's size is 4 mm, girdle the vine by removing a 4 mm wide ring of bark from the main stem and dip the clusters in 40 ppm GA_3 solution. While girdling care should be taken to remove whole of the bark without causing any injury to the wood. GA_3 solution is prepared by dissolving one gram of GA_3 in 50 ml of absolute ethyl alcohol (ethanol) and making the final volume of the solution to 25 litres.
3. After one week of the first GA_3 dipping repeat the dipping second time.
4. Keep the moisture of soil near field capacity throughout.
5. Harvest the crop when desired T.S.S . is attained in the berries.

Manures and Fertilizers

To the newly planted vines apply 20 gm of urea in March and repeat the dose in May and July to complete the growth up the bower in one season. From second year onward for vines planted at 3 × 3 m distance the following fertilizers doses and schedule may be adopted.

Age of the Vine (Years)	*Farm Yard Manure (kg)*	*Urea (46% N) (kg)*	*Superphosphate (16% P_2O_5) (kg)*	*Muriate of Potash (60% K_2O) (kg)*
1	20	0.4	1.5	0.25
2	35	0.5	2.5	0.35
3	50	0.6	3.5	0.50
4	65	0.8	4.0	0.65
5 and above	80	1.0	4.5	0.80

Apply whole farm yard Manure (FYM) and super phosphate as well as half N and half K fertilizer just after pruning. Apply remaining N and K after fruit set in April. Use of these doses have encouraged excessive vegetative growth on all vines irrespective of the system of training adopted, particularly in clay loam soils. The excessive growth causes barreness in vines, profuse growth causes shade in bower system and intermingling of long overgrowing branches. The hanging of shoots particularly in head system encourage fungal diseases. In such situations the nutrient contents of leaves and soil should be got tested and fertilizer dose suitably reduced/adjusted. If bearing is heavy in Perlette vines, give two sprays of urea @ 1.0 percent, first at full bloom and second at fruit set to get higher yield and better fruit quality.

Irrigation

The good growth of vines and berries on the clusters depends directly on the supply of irrigation through out the growing period of vines. The grape vines have fibrous root system which mostly spread in the top 20 cm of soil. A little depletion of soil moisture can cause diffusion pressure deficit within the vines, leading to growth inhibition. During dormancy vines can tolerate soil moisture stress for a sufficient, long time. To keep roots active and check the damage due to frost and low temperature in North India vines may be applied light irrigation if rain is not there during December or January. Following irrigation schedule may be followed for taking good crop from vines.

Time	*Number or Irrigation*
After pruning in February.	One irrigation
At the time of sprouting of buds.	One irrigation
After fruit set in April till-May.	4-5 days interval- 6 to 8 irrigation
15th May to fruit harvest.	Irrigate only when prolonged dry
July to October	spell is there.
November to January	One irrigation if soil get extremely dry in end December.

Give only light irrigation to the vines. No flooding of the whole vineyard should be done. The vineyard should be divided into small blocks depending upon the nature of soil. The block may have 4-8 vines. Flood with the irrigation a few days before the harvesting of grapes is carried out. Due care to irrigate the vines should be given even after the harvest of the fruit.

Weed Control

Vines are planted closely and normally on bower system, hence due to shade weeds could make little growth. But in vineyards trained on kniffin and head system the weeds get ample irrigation, nutrition and light to flourish. Therefore, it is very necessary to control the weeds. To check the emergence of weeds give pre-emergence application on Hexuron 30 WP (diuron) @ 1.2 kg/acre in the first week of March after thorough cultivation in February. To check the growth of perennial weeds spray gramoxone 24 WSC (paraquat) @ 6 ml/litre of water twice or Round Up 41 SL (Glyphosate) @ 8-10 ml/litre once, when the weeds have attained good height of 20-30 cm. These weedicides should be sprayed on a calm day to avoid drift to the foliage of vines.

Intercropping

No intercrop should be grown in vineyards as growing of intercrop shall be at the cost of training of vines. More over juvenile period of grape vine is short and vines are planted very close where no implement can move freely. However, in the first year some vegetable crops like peas, potato, turnip, radish or carrots can be grown. Cucurbits can be grown for first two years. To improve the organic mater content of the soil, leguminous crops like guara, cowpeas and senji can be grown-for green manuring. No Rabi season crop should be sown in vineyards as these crops do not require water to maturity in April when vines need it urgently.

Maturity and Harvesting

Grapes should be harvested only when all the berries have developed the unique colour and desired TSS of the cultivar.

Grapes are harvested by repeated pickings since the bunches do not ripen at one time. Taste is the most valuable indicator of the ripeness of the bunch. The berries at the shoulders ripe first followed by centre and tip of the clusters. There may be some bunches having unripe berries at the tip which may be harvested and unripe portion clipped for marketing. The bunches should be plucked with secateur very close to the canes. During harvesting, bunches should be placed in the baskets very gently. Care should be taken to save the natural bloom on the berries. After harvesting in the evening the bunches should be kept in shade for packing.

Post-harvest Handiling and Marketing

The unripe, small sunburnt, soft and damaged berries in the bunch should be trimmed. Bunches should be graded according to fruit maturity and size. Different grades should be separately packed in containers. For local marketing mulberry or bamboo baskets can be used. For distant markets, CFB boxes of 2-4 kg capacity should be used. In the cold storage maintain the temperature of – 2 to 0°C with relative humidity of 85 to 90 percent where it can be kept for 40 days. Shelf-life can be improved if heat of berries is immediately removed before actual storage by forcing the air through the boxes at 2°C less than cold storage temperature.

Special Problems

Barrenness

It has become the major problem of vineyards of North Indian plains. The vines have developed unproductive wood. At the time of pruning most of the shoots are found to be dry. A few fruiting canes per vine are available, hence reduced productivity. The canes produce bunches with few berries. It has been noticed that excessive vegetative growth of vines leads to over shading or immature elongated shoot growth resulting into failour of production of floral primordia development. There may be many reasons for this physiological state. Over irrigation, excessive nitrogen application, defective training, wrong pruning and keeping of high number of fruiting canes in early years of bearing. Both Perlette and Anab-e-Shahi, cultivars are prone to barrenness. Some workers feel that it is due to lack of flower bud formation due to increased diameter of canes. There is direct relationship with vigour of shoots and dryness.

Barrenness of vines can be prevented by following proper pruning practices and plant protection measures. Avoid taking heavy yield from the young vines. The excessive growth can be checked by spraying cycocel (CCC) @ 1000 ppm to 2000 ppm, Just after fruit harvest. Inclusion of Bordeaux mixture 2 : 2 : 250 spray during July-August to control fungal diseases can help in the reduction of barrenness.

Water Berries

Development of water berries in grape clusters is very common. These berries do not develop into full size, remain off coloured, lack normal sugar and have more acidity. The berries look like small cellophane bags half filled with sap and remain hanging on the clusters. The berries do not have firm pulp and shrivel or may dryup by the harvest time. The main reason is overcrowding of berries in bunch. Excessive vegetative growth at the cost of developing of berries in a cluster due to more nitrogenous fertilizer application and over irrigation in heavy soils is also responsible for the water berries formation. Water berry formation can be reduced by applying balanced dose of fertilizers and checking flooding of vines during irrigation.

Shot Berries

The presence of very small berries in the clusters of grape cultivars particularly perlette is very common. These berries are usually seedless and are called millerandage. Such berries develop due to poor pollination or poor carbohydrate accumulation after fertilization of flowers. Generally small berries drop due to embryo abortion or lack of nourishment. Those do not drop become shot berries. The treatment of clusters with GA_3 and thinning of the clusters help in checking the formation of shot berries.

Bud and Flower Drop

This physiological disorder is known as colore or shelling, which is a serious problem of Thompson seedless cultivar. The panicles become nacked. The shattering may be due to inadequate pollination/fertilization. Due to heavy shattering clusters become very loose or straggly. Grape cultivars differ widely in their response to shattering. Shattering can be prevented by following proper cultural practices and by spraying plant growth regulators to maintain canopy of vines. Exogenous applications of NAA or zinc before anthesis can reduce this problem.

Cluster Apex Wilt

The wilting of apex of the cluster is a big problem in Perlette and Thompson seedless cultivars. The apical portions of the clusters contain wilted berries at maturity. These berries remain acidic in taste. Sometimes whole of the apex dries up. This disorder is usually associated with compact bunches and low moisture in soil at development stage of the berries.

PLANT PROTECTION MEASURES

A. Insect-Pest

1. Grape Vine Thrip (*Rhipiphorothrips cruentatus*)

It is highly polyphagous. The nymphs and adults rasp the ventral surface of tender leaves and flower stalks and suck the oozing cell sap. As a result of their, intensive feeding the leaves develop silvery white scorchy patches with curly tips gradually get deformed and ultimately drop. Attack on flower stalks results in shedding of flowers. Berries are also attacked resulting in sub formation on the berries. The attack occurs during February-March and again in September.

Management

Spray Malathion 50 EC @ 1 ml/litre at panicle emergence and again just after fruit set.

2. Jassid (*Arboridia vinifertata*)

Nymphs and adults of hoppers usually suck the cell sap from ventral surface of leaves. The feeding spots on the leaves become pale. In case of severe infestation the affected leaves turn yellow, gradually start curling, become brown and ultimately drop. Its attack is serious after rains. It is also responsible for indirect loss by producing honey dew which serves as a substrate for the growth of sooty mould fungus on foliage and fruits. It affects the production of the grape vines and depreciate the quality of grapes.

Management

Spray sumithion (fenitrothion) @ 1 ml/litre or Hexavin 50WP (Carbaryl) @ 3g/litre of water.

3. Leaf Roller (*Sylepta lunalis*)

The eggs are laid on the ventral surface of leaves. On hatching, the young caterpillars feed on epidermis of leaves and skeletonize the same. Later these caterpillars roll the leaves and feed within. The pest is very active during monsoons.

Management

Spray Thiodan 35 EC (Endosulffan) @ 1.5 ml/litre of water or Hexavin 50 WP (carbaryl) @ 3g/litre of water as soon as the pest is seen.

4. Defoliate Beetles (*Adoretus spp.*)

Beetles appear on the onest of monsoon and feed on leaves during night and hide during the day. In case of severe attack, berries are also scrapped near the apical end. Eggs are laid in the soil, grubs feed on roots and can cause death of the vines.

Management

Spray in the evening Hexavin 50 WP (carbaryl) @ 3g/litre of water as soon as the damage starts. Repeat the spray after a week again.

5 Yellow and Red Wasps (*Polistes hebraeus and Vespa orientails*)

The Waps attack the ripe berries. The injured berries give bad look to the bunch and have to be removed.

Management

Burn or smoke the wasp nests present on nearby hedges/ trees at sun set. On a small scale damage can be checked by covering the clusters with muslin cloth.

6. White Ants

White ants attack the roots and climb the trunk of vines. They feed on the sap and wood under the scaly bark. Attack can be seen in periods of drought. White ants weaken the vines and low yields coupled with death of vines is the result of severe attack.

Management

Spray the vine trunks with Chloropyriphos @ 4 ml per litre as soon as the attack is noticed. Repeat the spray after a week again. The roots may also be treated with solution containing chloropyriphos @ 10 ml/litre. One litre of the solution per vine is enough to control white ant attack to roots.

B. Diseases

1. Anthracnose or Die Back (*Elsinoe ampelina*)

Small light brown spots appear on young leaves, which later enlarge, turn dark brown and give shot holes appearances. In severe attack vines get defoliated due to falling of attacked leaves. Dark-brown sunken spots with raised margins develop on new shoots/canes leading to their death from tip backwards. Similar spots appear on laterals of clusters and berries.

Control

(*i*) Prune the infected canes/shoots during January and spray Bordeaux mixture 2 : 2 : 250. Repeat the spray in March.

(*ii*) Spray Bavistin @ Ig/litre of water in last week of April.

(*iii*) Give another spary of Bordeaux mixture (2 : 2 : 250) in last week of May.

(*iv*)) Spray Bavistin I g/litre of water in mid July.

(*v*) Spray Bordeaux Mixture (2 : 2 : 250) in mid August. Spray should be made to fully bathe the foliage of the vineyard.

(*vi*) If need be give another spray of Bavistin @ 1g/litre in September.

2. Cercospora Leaf Spot

It is caused by fungs Cercospora sp. The necrotic small area on leaves with straw coloured centre and reddish brown margins appear.

Control

If the above recommended sprays are given to control Anthracnose, the disease may not appear at all. If at all it appears, the same schedule may be followed.

3. Downy Mildew (*Plasmopara viticola*)

Light yellow oily spots appear on the upper surface of leaves, which on the lower surface are covered with white downy growth of the pathogen. Later the spots become brown and brittle. Leaves with many active spots drop prematurely. The disease starts appearing in nursery and on grown up vines in March-April. After rainy season, it assume serious proportions and continues to appear in humid weather. It affects tendrils and fresh growth of shoots.

Control

The same schedule of sprays may be followed as per anthracnose.

4. Foot Rot or Collar Rot (*Rhizoctonia* Sp.)

In case of young vines the roots and collar region turn brownish with the shredding of bark and internal brown discolouration. Leaves turn yellow and ultimately the vines wilt and die.

Control

(i) The cutting before planting should be dipped into 0.2% Ziram suspension (2g per litre).

(ii) The soil in the pit should be drenched with Captan (400 g in 100 litres of water) thoroughly before planting the cuttings.

5. Powdery Mildew

Caused by uncinula necator. White Powdery growth of the fungus on leaves, tender branches and berries. The disease infects all aerial parts of vines. On leaves, white powdery patches appear which enlarge, coalesce and became dirty white at a later stage. The young leaves curl upwards and severely infected leaves are shed prematurely. In south western Punjab, it appears in the form of yellow, undetectable spots on the upper surface of the leaves and its presence is felt when it appears on the berries as dirty white growth.

Control

Spray the vines with 0.25% wettable sulphur (1.25kg in 500 litres of water) or spray Bayleton @ 40g or Topas 10 EC @ 40ml/100 litres of water in mid-March, end April and 10 day after again in May.

6. *Rotting of Berries*

(Black mould rot, blue mould rot, etc.) is caused by various kinds of air borne fungi, such as Botrytis, Rhizopus, Aspergillus, Pencillium yeast. Grape berries are attacked when still on vines. The wasps aid in injuring the berries and releasing the juice which serves as substrate for growth of fungi. In Perlette which has compact bunches, the growth pressure of barriers ruptures some berries and the released Juice flows into other berries, where the fungi grow.

Control

(*i*) For Perlette follow the practice of thinning of bunches as recommended for quality improvement.

(*ii*) Use insecticides or repellents to guard against wasps and other insects causing injuries to berries.

(*iii*) During June just before ripening spray the vines with 0.2% Ziram (1kg/500 litres of water). Stop spray a week before harvest.

BER

Botanical Name

Ber (*Zizyphus mauritiana L*)

Family

Rhamnaceae.

Ber has been a popular fruit of the rural India. In Punjab it has specific sanctity since some ber plants. *i.e.*, 'Baba Budha Ber', 'Dukhbhanjni ber' are still alive in Golden Temple at Amritsar. It is not fastidious to its soil and climatic requirements, hence, is grown through out India. Ber is known as poor man's fruit due to its wide cultivation. However, with the selection of high yielding cultivars with good quality fruits, it has become a commercial crop of Punjab and adjoining areas. Due to its hardy nature and low requirements for water during summer it is being cultivated in Rajasthan, Haryana and Uttar Pradesh on a large scale. In Punjab its cultivation is being taken up in the districts of Sangrur, Mansa, Bathinda, Ferozepur, Patiala and Roop Nagar. Ber is a rich source of vitamin C, proteins and minerals like calcium, phosphorus and iron. Ber is commonly known as poor man's apple, owing to its nutritional value.

Origin

Ber has been mentioned in ancient Sanskrit literature, it is said to be originated in India and China area.

Area and Production

India is the largest producer of ber. In Punjab ber has been planted on 2500 hectares, with annual production of 42847 MT.

Uses

Ber fruit contain 13 to 20 percent TSS and 0.2 to 1.0 percent acidity. It contains Vit. C (120 mg 100g of pulp), protein (0.94%), calcium (0.03%), and phosphorus (0.036%), iron (1. 14%), β-carotene (Vit. A) 80 ug/100g pulp also contain about 5.50% sucrose, 1.50 glucose and 2.0% fructose sugars. Ber trees are used for rearing lac insect, the leaves along with tender shoots are used as fodder, wood is used as fuel and for medicinal purposes. 'Ber fruit is usually eaten fresh. It can be used for the preparation of murabba, candy and chutney. Squash or nectar can be prepared from the pulp. The fruits of wild varieties particularly malah ber (Z. *numularia*) are dried in sun and are sold during off season.

Botany

Ber (*Zizyphus mauritiana*) belong to family Rhamnaceae. The tree is summer dormant. Usually a thorny shrub or a medium tree with 8-10 meter height. Leaves are alternate, simple, obtuse and broadly oval. Inflorescence axillary Cyme .Flowers bisexual and pentamerous. Ovary superior. Fruit is a drupe, may have greenish yellow to golden yellow colour of the epicarp edible portion is epicarp and mesocarp. Hard endocarp contain the seed.

The Chinese jujube (*Zizyphus jujube Mill.*) and mallah ber (*Z. numularia Burm F*) along with *Z.mistol* and *Z.xylopyrus* are other species of importance.

Climate and Soil

It is a very hardy fruit tree and perform well under adverse climate and poor soil conditions. It can successfully be grown in tropical and sub-tropical climate. Deep sandy to sandy loams are best suited for ber cultivation. However, good ber orchards are found in clay loam and some what alkaline soils too. The only requirement is assured irrigation during fruit development, *i.e.,* November to March. Frost damage the fruits but not the trees.

Important Cultivars

Umran

Trees are spreading and vigorous. Fruit is large and oval

with round apex. Colour of fruit yellow to golden yellow when fully ripened. Overripened fruits turn chocolate brown. Pulp sweet with 15 to 19 percent TSS. Fruit ripens in second fortnight of March. Tree is heavy yielder (150-200 kg). Fruits have very good keeping quality. Susceptible to powdery midew disease.

Sanour-2

This is one of the selections made from village Sanour in district Patiala. The tree is spreading and less vigorous than umran. TSS of pulp is 15-18 percent. Fruit ripens in second fortnight of March. Average yield 150 kg/tree. It is resistant to powdery mildew, disease and can be cultivated in humid areas.

Kaithli

The cultivar is selection from kaithal in Haryana. The trees are erect growing and vigorous than Sanour-2. Fruit is medium in size oblong in shape with a pointed apex. The colour of the fruit at ripening becomes greenish yellow. Pulp is soft with 18% TSS. Fruit ripen in end March. Average yield per tree is 100 kg. This is excellent table purpose ber fruit but sometimes can be attacked by powdery mildew.

Z.G. 2

The trees are spreading with drooping branches. The fruit is medium in size and roundish in shape. Very soft to eat. The colour of the fruit become light green at ripening. The pulp have excellent blend of acidity with TSS 13-15%. It ripens in end March. Average yield 150 kg per tree. Less susceptible to powdery mildew disease.

Wallaiti

Tree is semi-erect growing and vigorous growing. Fruit size medium to large and oval in shape. Epicarp light golden yellow in colour at ripening. Pulp soft and sweet with 15% TSS. It is an early ripening cultivar and ripens in March, Average yield 100kg per tree. It is also susceptible to powdery mildew. New ber varieties released are Goma Kirti, Thar Bhubharaj, Thar Sevika (Seb × Katha).

Propagation

Normally nursery men use the desi type of ber (Katha ber) stones for raising of the seedlings as the root stock but trials at Punjab Agricultural University with respect to rootstock have shown that ber should be propagated on *Zizyphus maurtiana* Dehradun. or Coimbatore. It is difficult to get the stones of these strains locally therefore the stones maybe purchased from Deharadun initially. It is recommended that mother plants of these rootstocks should be raised locally and entire demand for rootstocks be met from mother plants. Ber is a cross pollinated fruit crop therefore, raising of uniform seedlings is a difficult proposition.

The seed germinate with difficulty. The stones are as hard as peach. The stones are collected from ripe ber fruits. There are to methods of sowing of the seeds for raising of the rootstock seedlings.

In May sow the stones densely in a well prepared flat bed and cover the stones with sand + F.Y.M. mixture. Continue to apply irrigation to keep the bed moist. After 40-45 days of sowing, when the stones start splitting and plumules emerge. Pick up the seedlings at 2 to 4 leaf stage and carefully plant in lines in the nursery field. Apply water with a spray pump or sprinkler soon after the planting of seedings. This is a sure method of raising rootstock seedlings without any gap.

In May seeds can be directly sown in lines at 30 cm apart in the nursery field at a distance 10-15 from seed. Continue to apply irrigation at regular intervals. Weeding should be done once as soon as weeds emerge prior to the emerge, prior to the emergence of the seedling. When the seedlings become pencil thickness these are T-budded.

T-Budding

T-budding should be done during June to September at a height of 15 cm above the ground level. The lower most shoots/ sprouts emerging on the rootstock below the bud should be constantly removed after budding. After one month of budding,

50 percent of the tops may also be looped before removing the whole of top above the union points. The successful buds should be single stemmed soon after the sprouting.

Planting

The ber plants are normally lifted from the nursery along with whole of root system with sufficient soil during August to October for planting. The ber plants can also be planted bare rooted during December-January. For this whole root system is dug out carefully from 'watter' soil. The plants are given a little pruning to remove 50 percent of the vegetative growth. Bare rooted plants should be healthy and have at least 60 cm of height. Defoliation of plants after lifting from nursery is useful in establishment of plants in the field.

Before lifting the nursery plants, proper layout and digging/fillings of pits should be carried out. The layout may be done on square system of planting. To avoid attack of white ants the pits may applied 30g of lindane or 10 ml of chloropyriphos/liter of water per pit. Plants Should be placed in the centre of pits with the help of planting board. After planting, the sides of the earth ball/roots of bare rooted plants should be gentally pressed from all sides by standing keeping one foot on either side. Irrigation may be applied immediately after planting to settle the soil firmly. The planting distance for ber is 7.5m × 7.5m. However, for heavy soils and vigorous cultivars the planting may be done at 8.0 × 80m apart thus one requires 169 to 144 plants per hectare accordingly.

On sandy soils the root stock is 'T-budded *in situ*. This system is more successful than transplanting of budded plants in such a situation.

Care of the Young Plants

The newly-planted plants should be irrigated frequently at an interval of 7-10 days depending upon the day temperature. The sprouts appearing on the trunk below the bud union should be regularly removed with sharp Secateur. The young plants should be protected from the attack of insect pests.

Training and Pruning

Ber come into bearing within 2-3 years of planting in the orchard. The short juvenile period and vigorous growth of shoots compel the grower to provide proper training to the plants at an early stage to build a strong and viable scaffold system. It will be better to provide support to young plants to save breakage of bud union. The trunk should be kept straight. Revome the branches from ground level to 60-70cm of trunk height. Select 4-6 side branches on all directions at a sufficient distance (15-20 cm) from one another. The training should be completed within a year. In the second year top leader should be removed to keep an out growing lateral.

Pruning

Ber bears on new years growth in the axil of leaves. Therefore, to get optimum growth every year regular pruning is essential. The pruning should be done after 15th May. Pruning improves growth of the tree and fruit quality. If left unpruned for a number of years the ber trees become unproductive and laggy. It has been seen that unprund trees may have 10 metres long scaffolds and very little foliage and bear fruit at the tops only. It is essential to remove about 30 percent of the past years growth. Remove all the dead, dried and broken twigs. To provide more sunshine remove some of the secondaries too. Very severe pruning may be given after every 5-6 years of light pruning. For this remove 75 percent of the growth of secodaries and 50 percent of the major scaffolds. Apply Bordeaux paint on major cuts and Bordeaux paste on smaller cuts. While removing strong scaffolds or secondaries, care should be taken that bark should not get injured. Give a deep cut from down upward, then give cut from upper side coinciding the first cut.

Top Working

Inferior 'katha' ber trees can be top worked with superior cultivar. For this, head back the major limbs to keep 15 cm long stubs on one or two scaffolds. Remove rest of the mature wood judiciously during May. Many sprouts emerge on the stubs in July. T-budding should be done on 2-3 shoots on each stub. Remove rest of the sprouts. Continue to remove unwanted sprouts for a year. Save the sprouted buds from breakage by giving little pruning initially. The newly developed scaffold system from the sprouted buds come into bearing within two years.

Rejuvenation of Old Orchards

The ber trees if left unpruned for a number of years may become uneconomical. These trees produce little crop of small sized fruits. The trees also get infested with lac insect, powdery mildew etc. In such cases the trees can be rejuvenated. Remove the scaffolds with sharp saw by keeping 10-15 cm long stub. Keep only 4-6 stubs per trees during May. Select, out growing 2 sprouts on each stub during July-August same year. These sprouts grow very fast, hence, the pinching of tops should be done in August-September so that side branch my develop on these sprouts. Thus full grown tree is ready upto November-December of the same year. Give light pruning during May next year to get some fruit. Such trees can again yield heavy crops for another thirty years.

Manures and Fertilizers

Age (Years)	*Farm Yard Manure (kg)*	*Fertilizer Dose Urea (46% Ng/tree)*
1	20	100
2	40	200
3	60	300
4	80	500

Ber is summer dormant and pruning is done during summers. Therefore fertilization has to be done during rainy season. Apply whole farm yard manure just after removing the pruning wood in June. Urea should be split into two parts. One part should be applied in August and second part during November.

Irrigation

Water is life for every plant therefore, its use should be judiciously made to save water. No flooding should be carried out till plants cover whole of the space provided. First irrigation should be given in June after pruning for hoeing of the tree basins and for application of farm yard manure. No irrigation should be applied during rainy season. When there is drought apply irrigation after fruit set. Thereafter, give irrigation only at monthly interval up to February. Fruit start ripening from March onward and harvesting is almost complete in April under Punjab conditions. No irrigation should be given during and after fruit harvest.

Cultivation

Ber trees have short juvenile period. Plants come into bearing in the second or third year after planting. Therefore, deep ploughing by using discs harrow should not be done such a ploughing can damage most of the feeder root system. Give only light ploughing during June after removing pruning wood. Ber trees occupy most of the area due to their spreading nature therefore use of tractor in grown up trees should be avoided. The tree basins should be kept free of weeds by giving hoeing.

Intercropping

In the first 2-3 years of age, the plants occupy in little area and most of the land remain vacant. The vacant land should be judiciously used by growing compatible crops. No crop should be grown in kharif season since, it will need irrigation during dormancy of ber plants. Grow 'rabi' crops like wheat, gram, massar, barseem, sarson, toria, etc. For the first two years, later on grow only toria or winter vegetables so that the field is free during March-April for easy harvesting of the fruits. Exhaustive and tall growing crops like bajra or jawar should not be grown at all.

Weed Control

Most of the weeds remain under check due to summer

dormant nature of the ber tree. However, perennial weeds like baru grass, dubh grass, parthenium and puthkanda, etc. do raise their head during growing period of trees. To check the perennial weeds, spray Round Up (glyphosate) @ 10ml/litre or Gramoxone @ 6ml/litre when the weeds have attained sufficient vegetative growth (20-30cm height). Gramoxone can be sprayed in tree basins also. Choose a clam day for the spray to avoid spray drift to the ber foliage.

Control of Fruit Drop

Usually ber fruit does not drop physiologically, drop is normally due to pathological reasons. Sometime due to increase in atmospheric humidity and vegetative growth, there may be fruit drop in some ber cultivars in October-November. Single spray of 10-20 ppm Naphthalene acetic acid (NAA) one in mid-October shall be sufficient to control fruit drop.

Fruit Maturity and Harvesting

Ber fruit matures differentially, depending upon the time of fruit set. Some cultivars mature earlier than other. Umran is late ripening cultivar but Gurgaon Gola ripens early. Near Delhi and Agra ber ripens in February-March where as in Punjab it ripen in March-April. Fruit should be picked up at right stage of maturity. It should be picked when it has acquired normal size and characteristic colour of the cultivar. For example Umran fruit develop golden yellow colour at ripe stage. In all 4-5 pickings are usually done from mid March to mid-April in Punjab. No fruit should become over ripe on the tree, since the taste get deteriorated.

For uniform and early ripening of ber fruit, one can spray at colour break when some of the fruits on a tree have developed the colour with 400 ppm of ethephon (250 ml of ethephon in 300 litre of water). Ethephon advances ripening of ber fruits.

Grading and Packing of Fruits

After harvesting of ber fruits, grading should be done. Remove the undersized misshapened, cankered and wind damaged fruits. One can make 3-4 grades, *i.e.,* A grade best fruits of large size, B-grade-medium sized good coloured fruits, C-grade small sized well coloured fruits and D-grade poorly coloured fruits of all sizes and left outs. The highest percentage of fruits fall in B-category followed by A grade. C and D grade may have just equal quantities. A and B grade fruits are preferred in the market as well as by the consumers. A B and C grade fruits should be packed in corrugated fibre Board (CFB) cartoons, wooden crates, plastic crates, polynets or baskets of convenient sizes. D-Grade may be marketed in gunnybags.

Shelf Life

Umran fruits can be stored for three weeks in home refrigerator (7-8°C) in perforated polythene bags. The shelf- life of fruits can be enhanced up to 30 days if the fruits are treated with 6 percent wax-emulsion and packed in perforated polythene bags of 100 gauge thickness, before storing in the commercial cold storage (7-8°C) and 85-90 percent RH.

PLANT PROTECTION MEASURES

A. Insect-Pests

1. Fruitfly (*Carpomyia Vesuviana*)

It is one of the specific and serious pests of ber. Fruitfly deposits eggs in the epicarp of developing fruits. After hatching, the larvae enter into pulp and start feeding. The infested fruits become deformed, turn brown, rot and drop off. Fully grown maggots emerge out of the fruit by making a hole and pupate in the soil.

Management

(*i*) Clean cultivation/sanitation of orchards by picking and destroying the infested fruits should be done.

(*ii*) To escape egg laying on the fruits, harvest the fruits green and firm stage and do not allow the fruit to ripe on the tree.

(*iii*) Rack the soil around the trees during summer to expose the pupae to heat and natural enemies.

(*iv*) Spray Rogor 30EC (dimethoate) @ 2ml. of water during second and forth week of February. No spray should be given 15 days prior to fruit harvesting.

2. Leaf Eating Caterpillar (*Euproctis spp*)

The young caterpillars initially remain gregarious and scrap leaves and tender fruits. Later, instars disperse and devour leaves, fruits and tender shoots. Affected fruits become unmarketable.

Management

(*i*) At early and smaller scale, pluck the leaves with egg masses, and young caterpillars and destroy.

(*ii*) Spary Hexavin 50 WP (carbaryl) @ 3gm/litre of water as soon as the damage is noticed.

3. Lac Insect (*Kerria spp*)

Although useful if reared commercially for lac production, their presence on ber trees considered as harmful because they devitalize the trees and affect adversely the yield of fruits.

Management

(*i*) Remove and destroy the infested dry part and scrap off the infested twig before treatment.

(*ii*) Spray Rogor 30EC (dimethoate) @ 2ml. of water or Hostathion 40EC (triazophos) @ 2ml/1. of water.

B. Diseases

1. Powdery mildew caused by *Microsphera alphitoides zizyphi* young developing leaves and fruits are covered with whitish powdery mass of the causal fungus. This condition results in the premature defoliation and fruit drop. Fruits remain small and become cankered and cracked.

Control

Spary wettable Sulphur @ 2-3gm\L. of water or Karathane 40EC or Bayleton 25 WP @ 50ml/100 litres of water at flowering and then at fruit set. If disease exits spray again in November.

2. Leaf mould and Leaf spot caused *Isariopsis* and *Phoma macrostoma*. The under dide of the leaves shows a blackish and sooty appearance. Leaves turn yellow and drop prematurely. Nectrotic, small to medium *spots* appear on the affected leaves.

Control

Spray Bordeaux mixture (2 : 2 : 250) twice at an interval of 15 days or spray Blitox (copper oxychloride) @ 3g/L of water.

LITCHI

(*Litchi Chinensis*)

Family

Sapindaceae

Litchi is very delicious fruit and is very much liked by the consumer during summers. It is generally consumed as a fresh fruit. Litchi fruits has no competition from other fruits as it comes in the market in May-June and faces no marketing problem.

Origin

Litchi originated in southern China. It has been in cultivation in China for thousands of years. Litchi is very fastidious to soil and climatic requirements hence, the spread to other areas has been slow. In India litchi was brought from Burma in Bengal during 17th century. In Punjab, litchi was brought by Mahajan family in 1920 from Dehradum to Pathankot.

Area and Production

It occupies an area of 71.9 thousand hectare in India with annual production of 423.4 thousand M. tonnes. Bihar is the largest producer of litchi followed by West Bengal, Western U.P. and by Punjab. It is planted in submontanous tracts of northern and eastern India since, it likes mild climate with sufficient humidity. In Punjab it occupies in area of 1528 hectares only and produces 21262 MT.

Uses

Litchi fruit is mostly eaten fresh. Litchi squash is a refreshing drink prepared from litchi juice. Litchi juice is preserved for using it as such. Sun dried or oven dried fruits are known as litchi nut these are popular with Chinese. The whole fruit can be frozen for later use.

Litchi is a rich source of sugars and minerals like Calcium, Phosphorus and vitamins particularly riboflain (Vit B_2). The TSS may range between 15 to 20 percent.

Botany

Litchi is an evergreen tree belonging to Sapindacae family to which rambutan (*Nephelium lappaceum*) and Longan (*Euphoria longana*) belongs. The genus litchi has one more species which is found growing wild in Phillippines, extensively. *Litchi phillippinesis* can be used as rootstock for Litchi chinensis.

Litchi chinenesis has a medium to large tree with glossy leaves. Plants are slow growing initially and spreading in nature after first 7-9 years. Later on trees may grow as high as 12 meters. The bark is grey in colour. The leaves are lanceolate in shape some times curves backward (seedless cultivars). There are three types of flowers in a panicle. Male flowers open first. The second type, pistilate hermaphrodite flowers appear after the male flowers. Only these flowers bear fruit. Later on staminate, hermaphrodite flowers with united stigma open. The tree may continue to produce panicles for a month or so. The unsynchronized opening of flowers call for cross pollination. It is always useful to plant two to three cultivars in an orchard. The flowers are insect pollinated particularly house flies and honey bees do the job.

Flowering and Fruiting

In North India the panicle in litchi appear from February to end March. The flowers start opening from end February onward and continue for more than 20 days in a cultivar. Initially the seed develop and in the later stage of fruit development aril start developing at a faster rate than seed. At this stage when the aril develop fast the cracking of fruit may take place.

Litchi fruit is a one seeded nut. The fruit size depends on the cultivar usually 3 to 5 cm in diameter. The aril is covered by a thin leathery pericarp which may develop a peculiar pinkish red colour at fruit ripening stage depending on the cultivar. Aril develops and covers the whole seed. The thickness of the aril also varies from cultivar to cultivar similar to seed. The pulp is very juicy, sweet and soft in texture.

Climate

Litchi performs well under humid conditions in subtropical climate. The fruit size and other fruit development characters are greatly affected by the humidity in the atmosphere. Litchi prefers valley like situations but can be grown on higher altitude with sufficient moisture condition. Litchi is prone to severe frost damage. Severe frost in winter and harsh summers are limiting factors for the success in litchi cultivation. In North India litchi is performing very well from Dehradun to Gurdaspur belt, without any major problem. But in central districts along Grand Trunk road some fruit cracking may take place. It needs about 200 chilling hours to facilitate fruit bud differentiation. Summer temperature more than 42°C is harmful for litchi trees. Severe fruit cracking and sun burning of fruit take place in severe summers.

Soil

Litchi is very fastidious to its soil requirements. It thrives well on deep well drained loamy soils, with high organic matter content and free from any hard pan. Since *Mycorrhiza* develop on the roots very fast in little acidic soil reaction, hence, litchi plants grow fast in acidic soils. It can be planted in soils with 6.5 to 8.0 pH. The roots can tolerate submerged conditions for a number of days.

Recommended Cultivars

Dehradun

As the name shows this cultivar originated in Dehradun is being cultivated for its early ripening habit from Dehradun to Jammu areas of Northern India. It is a regular and heavy bearing cultivar. The trees are fast growing, vigorous, spreading and upright in nature. The new emerging foliage is light green in colour. The leaves become dark green at maturity. The flowering take place in February and full bloom in end March. Fruit ripens in June. Fruits develop attractive colour, but are prone to splitting. The pulp is sweet and juicy, medium soft in texture. Juice has 17 percent TSS and 0.48 percent acidity. The pulp stone ratios is 3.75:1.

Calcuttia

This cultivar has spreading umbrella type of trees. The new foliage is light rose coloured. The new leaves takes 7-10 days to become green. The mature leaves are broad from the center than Dehradun and show waves in the lamina. It is biennial in bearing. It bears very heavily in on year and little or no crop in off year. The panicles emerge in March and full bloom is around 13th April. The fruits ripens in end of June. The fruits are less prone to cracking. The pulp is soft and juicy with good flavour. The juice has 18 percent TSS and 0.47 percent acidity. The pulp stone ratio is 4.78 : 1.

Seedless Late

As the name shows this cultivar have fruits with shriveled seeds with sufficient thick aril. The trees are slow growing but spreading and have more juvenile period. The young emerging leaves are rosy in colour. The mature leaves curl backward. The shoots show rosetting type of foliage. It is also prone to irregular bearing. The fruits ripen in the end of June just along with Calcuttia. Fruits develop deep carmine red colour. The pulp is soft sweet, and juicy. The Juice has 18.7 percent TSS and 0.3 percent acidity. The pulp stone ratio is 28 : 1. The mature tree may yield about 80 to 90 kg of fruit.

Rose Scented

This cultivar has aroma in the fruits. The fruits skin has pink colour. The fruits are heart shaped with greyish white and sweet pulp. The pulp stone ratio is 6.4 : 1.

Saharanpur

This cultivar is synonym of Early large Red and Pinjor

Common. This is an early ripening cultivar and ripens in first week of June. The fruits are large in size, heart shaped and skin pink to deep orange in colour.

Muzzafarpur

This is an important cultivar of Bihar. It is a high yielding cultivar, may bear 80 to 100 kg of fruits per tree. The fruits resemble Saharanpur cultivar. This early ripening cultivar ripens in first week of May in Bihar hence, is less prone to splitting. This cultivar has good future in northern India, as here it ripens in mid-June. The TSS of Juice is 18 percent with 0.48 percent acidity. The pulp stone ratio is; 4.78 : 1 just as in Calcuttia. Other promising varieties are Swarna Rupa, Sabour Madhu, Sabour Priya.

Propagation

Vegetative Propagation

Litchi is commercially propagated through air layering. For this purpose, straight growing, one year old shoots are selected. A 3cm, ring of bark 30-40cm from the apex of shoots is removed during August. Moist sphagnum moss is wrapped around the ringed portion and a part of shoot. The moss is covered with polythene. The roots can be seen from the transparent polythene. The air layers (gootee) are removed from the parent plant during October and planted in the nursery for one to two years before actually planting in the field. After transplanting the air layers, the beds are kept in moist condition with light irrigation. While handling the air layers care should be taken to provide support of hand to the roots in the moss grass. The nursery beds should be provided sarkanda shelter during winter to protect the litchi plants from frost damage. For better transportation and easy handling during planting, the air layers can be planted in polythene bags. The polythene bags of size 20cm wide and 30cm in length should be filled with a mixture of FYM and soil in the ratio of 50 : 50. The air layers are planted in this mixture and kept moist. It is easy to keep, these bags under net or shade or under a big litchi tree to protect it from frost and high temperature in summer. The plants in bags get better care than in field. It is easy to handle bags than earth balls in the field also. The chance of death of the plant at planting time is also rare as compared to earth balls.

Planting and after Care

The planting of litchi should be preferred during September-October. However, in the lower hills where rains continue up to August end planting can be advanced to end July. Two years old plants in the nursery should be preferred over one year old. The mild temperature with humidity during October and November help in the establishment of litchi plants better than those planted in July-August. Plants should not be planted too deep in the pits. Deep planted plants do not perform well. Litchi is generally planted at 8 to 9 metres apart in a square system. Thus 144-121 plants can be accommodated in one hectare. For planting on hexagonal system litchi should be planted at 10m × 10m, which require 105 plants per hectare. The pits should be prepared two months prior to planting. One metre deep and one metre diameter pits should be dug. Keep top half of soil on one side and lower half on the other. Mix farm yard manure 25kg well with the top soil and fill the pits 5cm above soil level. Apply two litres of Dursban 20EC (chloropyriphos) solution @ 10ml/litre of water to the filled pit. Apply irrigation to settle down the soil in pit. At the time of planting lift some soil from the centre of the pit with the help of a planting board and place the earth ball in the hole. Level the soil around the plant. It is preferable to use soil from a litchi orchard for leveling of the hole of newly-planted plant. This will help in the fast growth of newly-planted plants due to the presence of mycorrhizal fungi. Apply irrigation soon after planting.

Wind Break

It is very essential to save the litchi plants from the high velocity hot winds which can break the litchi plants or desiccate the foliage. So, plant a wind break by planting Jamun, Mango seeding or Arjuna interplanted with mulberry and other shrubs like bouganvillea around the litchi orchards two to three years ahead of litchi planting. A 1.5m deep trench should be dug at 1.5 to 2.0m away from the wind break so that roots of wind break do not compete with litchi roots for nutrition.

Protection to Young Plant

The establishment of litchi orchard is little difficult job. The mortality of newly-planted is so high in the early years that one can see a few established plants in an orchard. Farmers start growing annual crops again. For saving the young plants mulching should be done immediately after planting. White ants attack should be checked by applying chloropyriphos solution-twice @ 10ml/litre in November and then in March every year for three years. For producing shade and checking the effect of high velocity winds grow 'arhar' all around the plants at a distance of 1m. The same 'arhar' plants can be used to provide shelter during winter. The southern side of the basin should be kept open by uprooting some arhar plants. Arhar may be uprooted in the third year and if needed the re-sowing may be carried out at a distance of to 1.5m so that arthar roots may not interfere with litchi roots.

Training and Pruning

No prunings is done for the first two years. Remove only the lower most branches upto height of 50 cm. The removal of branches should be staggered through out for 3-4 years after planting. Keep sufficient number of leaves in the first year so that plants have profuse growth. Do not allow any branch to criss-cross or to grow inward.

Litchi plants take the shape of an umbrella of its own but very slowly. Training should be done on modified leades system of training which is completed after 3rd or 4th year of planting. Trees come into bearing in 5th year of planting depending on the cultivar and plant vigour. New growth automatically take place when a small part of shoot bearing fruit pancile is also removed during fruit harvesting.

Rejuvenation

When trees become too old and branches start intermingling, the yield decreases due to over crowding and lack of aeration and sun shine. At this stage the trees should be rejuvenated. The main scaffolds are headed back to keep 30cm of stubs during January just like in mango and pear. Keep only 5-6 scaffold stubs per tree. Top most scaffolds should be removed to bring the height of the rejuvenated tree down. Many sprouts come up on the stubs. Select 2-3 out growing shoots per stub. Thus there will be 10-12 branches per tree. The profuse growth in the first year it self-produces healthy trees again within 2-3 years of rejuvenation. These trees can yield good crop for a number of years again.

Irrigation

Sufficient soil moisture is a prerequisite for profuse growth and good fruiting in litichi. During rains no irrigation may be given. Shortage of moisture in soil adversely affect the growth of trees. Young as well as bearing trees need frequent irrigations. The plants should be irrigated after 4-5 days during hot months and interval may be increased to 7-10 days as the temperature falls.

No irrigation should be given in October and November. From December to February irrigation may be given at an internal of 20-30 days depending upon the winter rains. No irrigation is given at full bloom stage which normally comes in mid March to mid April depending upon the cultivar under Punjab conditions. During the fruit development period, *i.e.,* May and June irrigation may be given at an interval of 4-5 days. Sufficient moisture in soil reduce the cracking of fruit. Young trees should be irrigated through modified basin system of irrigation. The basin should be gradually increased as the plants spread. Mature bearing trees should be irrigated through flooding of small segments. Each segment may including 4-6 trees. In such a way sufficient water will be made available to the trees. Much of the water will be saved which otherwise would be needed for flooding of an acre, hectare or more as one segment. Growing of 'moth', cowpeas as cover crop in basins shall help in the fast growth of litchi plants.

Manuring and Fertilization

Litchi plants require sufficient organic matter for their roots to spread and make good growth of the foliage. The early profuse growth is required to reduce the juvenile period and for getting high early fruit yield. To obtain and maintain this growth sufficient N, P and K is required as nutrition. Nitrogen deficient plants grow slowly and have light green foliage which may turn yellow soon. Old leaves show yellowing earlier than the new ones. Phosphorus and potassium deficient plants show stunting of plant and scorching of old leaves. A healthy litchi plant should take 3-4 flushes in a year. Following fertilizer schedule is being recommended.

B. Fertilizer Schedule for Litchi

Age of tree (Years)	*Farm Yard Manure (kg)*	*Urea (46% N) (g)*	*Superphosphate (16% P_2O_5) (g)*	*Muriate of Potash (60% K_2O)(g)*
1 – 3	10 – 30	100 – 300	150 – 450	50 – 150
4 – 6	40 – 60	400 – 600	600-900	200 – 300
7 – 10	70 – 100	700 – 1000	1050 – 1500	350 – 500
Above 10	100	1000	1500	500

Farm yard manure, super phosphate and Muriate of Potash should be applied in December under the tree canopy uniformly and should be mixed well with soil. Urea should be split into three equal parts for the first three years of age. After that it may be applied in two equal parts. Apply urea in February; April and June to the young plants and in February and April to the bearing plants. In off year, non-bearing plants may not be given phosphorous and potash fertilizers and quantity of urea should also be halved.

Mircronutrient Deficiencies

Deficiency of Zinc, Iron and Manganese has been noticed in both young and bearing litchi plants. Therefore, in addition to the above fertilization, mixture of Zinc Sulphate plus Ferrous Sulphate @ 2 gm/litre should be sprayed during April. In May Zinc Sulphate + Manganese Sulphate should be sprayed @ 2gm/ litre of water. These chemicals should be of laboratory grade. Such a light spray shall not bring other mircronutrient deficiencies.

Intercropping

Litchi takes 6-7 years to yield a commercial crop. Moreover due to slow growth and more planting distance, sufficient vacant space is available for the first 3-4 year. This vacant space should be judiciously utilized by growing intercrops. Fruit plants with less juvenile period as kinnow, guava, peach or plum can be planted as fillers in litchi orchards. Vegetables can also be profitably grown for 3-4 years. Pulses add nitrogen to the soil and give high returns. These should be preferred over growing of wheat or fodder crops. The fillers should be removed before these start interfering with the litchi plants. The nutritional requirements of crops, vegetables and filler plants should be separately attended to. If wheat is grown then keep a separate irrigation system for litchi plants so that plants should not suffer during March-April for want of irrigation.

Fruit Drop

It seems, fruit drop in litchi occurs due to failure of fertilization of the ovule since the drop occurs just after fruit set. The maximum drop has been recorded in Dehradun followed by Muzzaffurpur and Calcuttia. The micronutrient sprays recommended effectively reduced, the fruit drop in Dehradun cultivar. Growth regulator like GA_3 @ 40 ppm spray at full bloom stage can check fruit drop.

Harvesting and Handling of Fruits

The litchi plants start bearing after fifth year of planting and the yield continue to increase with the increase in tree size up to 20 years of age. Litchi cultivars start ripening of fruit as early as June as in Dehradun and continue up to July as in Calcuttia. The fully mature and ripe fruits should be harvested since no improvement in fruit quality take place after the fruit is harvested. The fruit should be harvested when a representative colour on the epicarp (Skin) characteristic of the cultivar has developed and finely pointed tubercles on the skin have flattened/smoothened.

While harvesting whole of the cluster along with 15-20cm of leafy branch should be cut with a sharp secateur. The presence of foliage will help the fruit to escape wilting. Three to four pickings are made per tree to fully harvest the trees since all the clusters do not ripen at one time. The harvested fruit is kept in shade to maintain freshness.

Grading, Packing and Marketing

Litchi clusters should be graded as excellent, poor clusters and single fruits separately. The packing is done in small baskets by placing sufficient foliage under the clusters and covering with newspaper and litchi leaves for local market. However, for distant market small wooden boxes are used to pack the litchi fruit. For fancy litchi fruits can be packed in 500 gm transparent packs without using any foliage or paper as cushion.

Since, litchi fruit looses its freshness soon after harvest thus the packed material should be transported to distant market as soon as possible. During transit avoid crushing of fruit in packs. Growers can earn upto rupees 70,000 per acre if marketing is properly done.

Storage

Normally litchi is consumed as fresh fruit. At ordinary temperature it can not be kept for more than two days. After that the epicarp looses its original colour become hard just like shell of an egg. The fruit can be stored for a few days in the fresh form. The most important character is to retain the colour of skin in the cold store. The phased marketing of the fruit form cold store shall add more income to the grower since, no other fruit can replace litchi and mango. Litchi fruit treated with 6 percent wax emulsion can be stored in perforated polyethylene bags of 100 gauge thickness in a commercial cold storage (2.2°-3.3°C) and RH 90-95 percent for one month.

PLANT PROTECTION

A. Insect-Pests

1. Leaf Roller *(Platypeplus aprobola* Meyer*)*

The new flushes are attacked by this pest through out the litchi growing area. The green coloured caterpillar rolls the leaf and feeds on lamina within the roll. Plants belonging to genus *Hibiscus* act as main host plant therefore, such plants should not be grown near litchi orchards. For initial control collect and destroy the rooled leaves as soon as rolling is observed on young plants.

Management

Spray Thiodan 35EC (Endosulphan) or Dursban 20EC (chloropyriphos) @ 2m/litre as soon as the attack is seen particularly in August-September. Every new flush should be observed to control the pest.

2. Litchi Nut Borer (*Blastobasis* spp.)

It damages the fruit by entering through a small pin head size hole near the attachment of peduncle and feed inside it. Affected fruits become unfit for consumption.

Management

(*i*) Clean cultivate the orchard, Remove the fallen fruits (mummies) from the orchard. Ploughing destroys the carryover of pests to the next crop.

(*ii*) Spray sumicidin 20 EC (fenvelerate) @ 1½ ml/litre of Dursban 20EC (chloropyriphos) or Thiodan 35EC (endosulphan) or Ekalux (Quinalphos) @ 2ml/litre 20 days after fruit set. Repeat second and third spray after 15 days interval with a different insecticide each time.

3. Grey Weevil

A small grey weevil feeds on the leaf edges giving 'U'

shaped cuts. The weevil remains active throughout the flushing period. The larvae feed on roots which could be checked by deep hoeing before panicle emergence.

Management

Spray Sevin @ 3g/litre of water or Thiodan 35 EC (Endosulphan) or Dursban 20EC (chloropyriphos) @ 2ml/litre as soon as U shaped cuts are seen on the foliage. Repeat the spray after 15 days to fully control the pest.

4. Bark Eating Caterpillar *(Indarbela tetraonis)*

It is a serious pest in neglected orchards of litchi. Only larvae after hatching make holes in scaffold crotches, they feed on surface of bark. The faeces and brass of wood hangs down at the holes. The girdling caused by the feeding of bark can be injurious to trees.

Management

Clear the holes with a wire, manually and inject kerosene oil or chloropyriphos solution 50 : 50 and plug the holes with mud.

5. Erionose Mite (*Aceria litchi keifer*)

Both adults and nymphs infest the leaves. Mites feed on the under surface of the leaves. They puncture and lacerate the tissues of the leaves with their stout rostrum and suck cell sap. Such leaves become thick, curl up forming hollow cylinders and finally dry. Young plants and nursery is severely attacked. Mites usually attack in March and maximum activity has been observed in July-August.

Management

(*i*) Spray wettable sulphur 4g\litre of water after harvest of fruit in July and again in January.

(*ii*) Spray Rogor (Dimethoate 35 EC) @ 2ml/L before and after flowering.

6. Citrus Mealy Bug (*Planococcus citri*)

Mealy bugs cause damage during flowering and fruiting stages, *i.e.,* February to April. Large number of nymphs crawl up the trees and congregate on the panicles or young shoots and suck the cell sap.

Management

Wrap the tree trunks with a sheet of alkathene from under the ground up to 50 cm of height. The lower ends of the alkathene should be burried in the soil so that no nymph could climb up the trunk. The upper end of the sheet should also be tied tightly. Kill all the nymphs falling/collecting on the ground physically or by spraying Methyl Parathion 50% @ 2ml/per litre of water at panicle emergence and after fruit set if mealy bug is observed.

Biological Management

Biological suppression of mealy bugs is an ideal technique. Beetle *Cryptolarmus montrouzieri* is a good predator to suppress mealy bugs on fruit crops. The predator can be had from Director of Biological Control (ICAR) farm post, Bellary Road, Bangalore (Karnataka).

7. Snails

Snails do not belong to class insecta but has become major pest of litchi orchards. It attacks the trunks and scaffolds and suck the cell sap. It is very active during rainy season (July-August).

Management

Apply furadaon 10g per tree. Mix the pesticide well in the soil near the tree trunks as soon as the snails appear in the orchard soil.

Bats and Birds

They cause considerable loss of fruit. The beating of drums or firing of crackers or by using bright strips, bats and birds can be scared.

B. Diseases

1. Fruit Rot

Fruit rotting at the time of fruit harvest has been observed in some orchards. Rotted fruit cause rotting in other healthy fruits. The marketing of fruits is adversely affected.

Control

Spray Bavistin 1gm/litre of water one week before harvesting of fruit.

2. Shoot Drying

The antharacnose attack the foliage and twigs which start drying.

Control

Spray Bordeaux mixture (2 : 2 : 250) in January and July twice a year to control shoot drying.

3. Lichens

Due to closer plantation than the recommended planting distance, the lichens invade the shaded foliage. This affects the photosynthetic activity of the trees. The yield decline as the lichen attack increases.

Control

Spray Bordeaux mixture 2 : 2 : 250 twice one in February and then August.

Physiological Disorder

The splitting or cracking of litchi fruit during development particularly at high atmospheric temperature is common to all litchi cultivars. As the summer temperature rises above 38°C fruit start cracking. In Dehradun high percentage of fruits crack just before harvesting. Rains at maturity and fluctuation in Soil moisture and atmospheric temperature greatly affect the fruit splitting intensity in litchi.

Control

(*i*) Irrigate the orchard at regular intervals after 15th of April so that soil moisture usually remain at field capacity up to harvest of fruit.

(*ii*) Spray GA_3 @ 100 ppm, when fruits have attained 50 percent of its size.

PAPAYA

Botanical Name

Carica papaya L.

Family

Caricaceae

Papaya is the most preferred fruit for kitchen gardens. Normally it is planted as a filler plant in orchards of fruit trees with long juvenile period. Papaya plants come into bearing within a year of planting.

Origin

Papaya originated in tropical America. Portuguese introduced papaya in India from Malacca during 1611. At present it is cultivated throughout the world in tropical and sub-tropical climate. It do not tolerate freezing temperature and frost.

Area and Production

The papaya plant has short life, hence the area under cultivation varies greatly in different years. In India it is cultivated over an area of 97.7 thousand hectares with annual production of 3628.9 thousand MT. In Punjab it is being grown as filler plant in mango and litchi orchards. No data is available for production/average.

Uses

Papaya is eaten as fresh fruit. It is a rich source of vitamin A (2020 IU per 100 g) and sufficient vitamin C (40 mg/100g). It also contains sufficient amount of Calcium and other minerals. The yellow colour of the pulp is due to caricaxanthin. Papaya contains vitamin B_2, 100 gm of pulp riboflavin 250 mg 100 gm of pulp, 0.6% proteins and 9% of carbohydrates. It contains papain a protein digesting enzyme. Papain is prepared by collecting latex from the unripe papaya fruits. Unripe fruits can be used as vegetable. Ripe fruits are used for making of Jam, Jelly, and Ice-Cream, etc.

Botany

Four genera belong to family caricaceae. Only *Carica papaya* produces edible fruits in genera *Carica*. Papaya plant is dioecious in nature but hermaphrodite forms also occur. It is dicot with straight, and soft stem. Plant is usually unbranched but branches develop where apex get damaged. Leaves are long stalked and fan like. Papaya fruit is climacteric in nature. Fruit is usually berry containing no or many seeds depending on the cultivar and climate. Shape of fruit also varies from oblong to spherical.

Climate

It requires warm and humid climate without frost. Low temperature affect the fruit and plant growth. Both foliage and fruit get damaged near 0°C. Hence, plants have to be protected by covering with gunny bags under North Indian conditions. It is also very sensitive to water logged conditions.

2 Soil

Papaya as a filler is grown from plains to sub mountains areas. It like soils which are well drained, rich in organic matter with near normal pH. Water stagnation for few days can be harmful.

Cultivars/Varieties

Pusa Delicious

A selection from progenies of variety 'Ranchi', its plants are gynodioecious, *i.e.*, both pistillate and hermaphrodite and medium in size. Plant start fruiting at a height of 1.10 meter from ground level. The plants may grow to a height of 2 m. The fruits size is large with oblong to oval in shape. Flesh deep orange with 10% TSS. Average yield per plant 40 kg.

Punjab Sweet

The plants are dioecious and may grow up to 1.90 meter

high. Fruit bearing starts at a height of 1.0 m from ground. Fruits are of large size and oblong in shape with pointed tip. Flesh colour is deep yellow. TSS of pulp ranges between 9.0 to 10.5 percent. The average yield per plant 50kg. It is least susceptible to citrus mite (*Entetranychus orientalis*)

Honey Dew

This variety is suited for North Indian conditions. The plants are of medium height and bear fruit heavily on the trunk. Plants are dioecious in nature, but the proportion of male plants is low. The fruits are large in size, elongated in shape with few or no seed. Flesh is sweet (TSS 8.0%) with pleasant flavour. This cultivar is also known as Madhu Bindu.

Coorg Honey Dew

This is a selection from Honey Dew. The plant is tall and bears heavily. Most of the plants are hermaphrodite. A few may produce only pistillate flowers. The fruits are large, variable in shape. The flesh is thicker but TSS (6.0%) is less than Honey Dew.

Co-I

The plants are unisexual with male and female plants occurring separately. The plant is of medium height and start bearing low on the trunk. The fruit shape is round with golden yellow skin and orange coloured flesh. Fruits are medium in size. TSS of pulp is 7%. Average yield per plant is 40kg.

Pusa Nanha

This is mutant of a local type cultivar developed at IARI. The plants are dioecious in nature. The plants are dwarf and grow up to 1.06m height. Average yield per plant is 35 kg. This cultivar is preferred for close plantation.

Pusa Giant

Plants are of dioecious in nature. Plants can with stand strong winds. Plants start bearing at 1.0m height. The plants come into bearing after a year. It bears large fruits weighing over 2 kg per fruit. The fruits are more suited to vegetable making and canning industry.

Propagation

Papaya is propagated through seeds. Seeds are extracted from ripe fruits and are rubbed with little sand to remove the white layer in which seed is enclosed. The seeds are washed and air dried before packing in polythene bags and stored at room temperature before sowing. The seeds loose viability in about 45 days, so due care should be given before extraction of seeds. Seeds can be sown any time from March to September. However, due to severe winter in North India, the seeds should be sown during July to September, when the seedling are of 20-30 cm in height and can easily be protected from frost and freeze of winter.

Prepare a mixture of farm yard manure plus soil and sand in equal proportions. Use this mixture to fill the polythene bags of size 25 × 10 cm having 8 to 10 small holes on lower part of the bag for drainage of excess water. Sow two to three seeds per bag 2-3 cm deep. It will be better if seeds are treated with 3g of Captan per kg of seed before sowing. After the emergence of seedlings drench them with Captan solution @ 2 g per litre of water to prevent damping off disease. Remove the extra seedlings from the bag. Each bag should have only one healthy seedling. The plants become ready for transplanting in the field within 40-50 days.

Planting

Pits of 50cm × 50cm × 50cm should be dug one month before transplanting is intended to be carried out. The planting is done as filler or at 1.5m × 1.5m apart. Fill the pits with an equal mixture of well rotten farm yard manure and soil. To control white ants add 30g lindane 5% dust while filling the pits. Transplant 2-3 seedling 15 cm apart per pit. While planting the polythene bags should be removed. Apply light irrigation. Uproot the extra male plants. Keep only one plant per pit when these have started flowering. Keep about 10 percent male plant population scattered throughout the orchard. There will be 4356 plants per hectare.

Intercropping

Papaya is itself planted as filler in mango, litchi and guava orchards because, it stays in the field for 2-3 years only. However if papaya is planted enblock then shallow rooted vegetables like onion, garlic, turnip, cauliflower can be grown for the first year. No crop should be grown after one year. Intercrop should not compete for nutrition with papaya plants.

Irrigation

Papaya is very sensitive to water logged conditions. Avoid flooding periodically. Divide the field into segments. Apply light irrigation at 6-7 days interval in summer and 15 to 20 days interval in winter depending on soil condition. If rain is there avoid irrigation. Water should not stay near the stem of plants to prevent collor rot.

Manuring and Fertilization

Papaya needs more nutrition due to its short juvenile period. It prefers soils with good organic matter. Apply 20kg of farm yard manure by mixing with 200g of urea 400g of muriate of potash in February or March every year per plant .A very high dose of fertilizers have been recommended in other states which do not hold good under Punjab conditions. The male plants should not be fertilized at all. The manure and fertilizer mixture should be mixed within a radius of 40cm around the plants.

Weed Control

Weeding should be done regularly to keep the basins free of weeds when the papaya is planted as a filler in orchards. Two hoeings one in February-March and other in July-August are sufficient to check the growth of weeds. No weedicides should be sprayed since papaya is a shallow rooted fruit crop and plants can be damaged.

Frost Protection

In North India, frost usually occurs in the months of December and January. Sometime late frost may occur in February also. Papaya is very sensitive to frost due to broad leaves and succulent stem. To save the young papaya plants from winter, the plantation should be done during October or November so that plants could be provided protection easily. Cover the plants with Sarkanda thatches or by pegging three support around the plants and covering with white polythene bags of the size of plants in first winter. Care should be taken that bags do not touch the foliage. Few holes may be provided in upper portion of bags for aeration. In the second year thick gunny bags are wrapped around the fruits and apex during winter season.

Harvesting and Fruit Handling

Papaya plants tend to over bear. Fruits are so crowded that they do not get proper space for development. Fruit thinning should be done to get well sized fruits. Proper size is attained after 5-6 months of flowering. Near ripening fruits change colour from green to yellowish green. Fruits should be harvested at maturity. After harvesting the fruit should be placed in single layer and covered with straw till ripening. For distant markets it should be packed in baskets by placing straw below and newspaper along with gunny bag above to avoid bruising of fruits. Papaya is a climacteric fruit hence, it can be artificially ripened by dipping fruits in 500 ppm ethephon. At ripening fruits attain golden yellow colour of skin.

PLANT PROTECTION

(A) Insect-Pests

1. *Aphid (Aphis gossypii* and *Myzus persicae*)

The young plants are attacked by the nymphs and adults. They suck the cell sap and act as vector of papaya mosaic virus.

Management

(*i*) Spray Nuvacuron (monocrotophos) 36 SL or Malathion 50 EC @ 1 ml/litre before virus attack, as soon as the young insects are seen.

(*ii*) Uproot and destroy the virus infected plants.

2 White Fly (*Bemisia tabaci*)

Both nymphs and adults damage the new foliage. Tiny white scale like objects cluster in between the veins on the ventral surface of leaves and suck the sap from the lamina. Leaves become yellow and curl downward. Leaves shed early.

Management

Spray Nuvacuron (monocrotophos) 36 SL or Malathion 50 EC @1 ml/litre of water.

Red Spider Mite (*Tetranychus cinnabarinus*)

The mites suck sap from ventral surface of leaves under silken webs. The yellow spots appear at the attacked portion and leaves abscise.

Management

(*i*) Remove infected leaves and dump them in deep pits or burn them immediately after removal.

(*ii*) Spray the plants with Rogor 35EC (Dimethoate) @ 2 ml/litre. Repeat the spray after 10 days.

Diseases

1 Anthracnose (*Colletotrichum gloeosporioides*)

Fruits at all stages of growth are attacked by the fungus. Pinkish masses of the spores of the fungus develop on the depressed portions of the fruits. In severe cases fruits get completely rotten and drop down. The infected leaves bearing spots show shredded margins. Leaves give characteristic shot hole appearance.

Control

(*i*) Remove the infected fruits and dump deep in soil.

(*ii*) Spray Ziram or Captan or Indofil M – 45 @ 2g/litre of water and repeat the spray after 14 days.

2 Collar Rot and Stem Rot (*Pythium aphanidermatum and Phytophthora palmivora*)

Roots and Stem of the plant rot producing wet texture. The yellowing of leaves and stunting of the plant growth results due to fungal attack leading to poor fruit development and finally death of the plant.

Control

(*i*) Improve drainage apply light irrigation.

(*ii*) Uproot and destroy the infected plants immediately.

3 Powdery Mildew (*Oidium caricae*)

It is a very serious disease of Papaya. Fungus grows on the under surface of leaves as whitish powdery mass. Yellow blotchs develop on the upper side of leaves. Fungus also infects the fruits.

Control

(*i*) Spray wettable sulphur 3g/litre of water.

(*ii*) Spray Bavistin @ 1.0 g/litre after 10 days of Wettable Sulphur Spray.

4. Papaya Mosaic (Virus)

Leaves become small show patches of green tissue with yellow background. The patches resemble blisters. The leaves curl and become wrinkled.

Control

(*i*) Uproot and destroy the infected plants.

(*ii*) Control the vectors, *viz.*, Aphids and White flies before the appearance of virus.

5. Papaya Leaf Curl (Virus)

Young leaves at the apex of the infected plants curl, twist and become deformed. The plants become stunted and bear no fruit.

Control

Same as in Papaya mosaic virus.

To avoid and damage following spray schedule may be followed.

(*i*) Spray Monocrotophos @ 1ml/litre in the last week to February or first week of March.

(*ii*) Spray Monocrotophos @ 1ml/litre of water in first week of July.

(*iii*) Spray Monocrotophos or Malathion @ 1 ml/litre in mid-July.

PHALSA

Bot. Name

Grewia asiatica or *subinaequalis*

Family

Tiliace

Phalsa is indigenous to India and is known with different names in East and North India. Very hardy fruit, hence can be planted on marginal soils, where rest of the fruits can not be grown. Its cultivation is more popular in the vicinity of cities/ towns. Phalsa fruit is liked to its acidic taste, coloured squash and syrup.

Origin

Phalsa has been mentioned in our Vedic books and is native to India.

Area and Production

It is minor fruit and is being cultivated on very small scale in each state. However, in Punjab, Haryana and Uttar Pradesh it is cultivated near cities commercially. In Punjab area under Phalsa is only 30 hectares with annual production of 196 tonnes approximately.

Uses

The ripe Phalsa fruits are purple in colour at maturity and may turn black at ripening on the bushes. Phalsa is considered to be antioxidant in nature due to its coloured fruit and juice. Whole fruit is eaten along with the seeds. The mildly acidic fruits are rich in Vitamin A, C and minerals. Fruit contain 55 percent juice with 11-12% TSS and 3% acidity. Ripe fruits are consumed as fresh. Juice and squash is relished in summers. In Unani system of medicines it is recommended to cure various ailments. Phalsa roots are used to cure rheumatism. The shoot can be used to prepare baskets.

Botany

Phalsa belongs to genus Grewia. It is deciduous bush in North India. It can be trained both as a bush and single stemmed small tree. In South India it is evergreen. It performs better in areas with distrinct winters. The stem is hard and brittle. The bark has mucilaginous juice, which is used to purifying to sugar or jaggery. Leaves are broad rough and light green in colour, with hairs on both sides. Flowers small, yellow in colour, appear in clusters on peduncles in leaf axis. Flowers appear in April in north India. Fruit ripens in June and is drupe with one or two hard seeds.

Climate and Soil

It prefers hot and dry environment during fruiting. In winter it goes dormant and shed its leaves. It resprouts in March. High temperature of June help in fruit ripening. It is not fastidious to its soil requirement. It can very easily be grown on poor soils. Loam soils are considered best. It makes good growth under scanty irrigation conditions.

Cultivars

No distinct varieties have been developed. Even if plants are raised from seed and flowers are cross pollinated, no varieties have come up. Two types tall and dwarf did not have much difference when planted in the same situation.

Propagation

Phalsa is commonly propagated through seed. Cuttings are difficult to root due to the presence of mucilage. Some reports say that IBA treated cuttings can root when planted in January. Bold seeds give 90 percent germination during July. Sow seeds on raised beds 2cm deep in lines 10cm apart. Seed to seed distance should be 2cm. Cover the seeds with mixture of sand + F.Y.M 50 : 50 ratio. Apply water with sprinkler, immediately after sowing. Avoid flooding of seed beds, failing which root rot fungus *Pythium* may appear. Apply 1% Bavistin solution after the seed germinate. Apply the solution Dursban 20EC (chlorophyriphos) @ 10ml/L of water after 30days of seed sowing to check the attack of white ants. Seedlings become ready for transplanting in January.

Planting

Transplant only healthy seedlings bare-rooted in the prepared pits in January. Planting is done at 1.0 × 1.5m apart in lines. A month prior to actual planting pits of size 0.5 metre deep and same diameter are prepared. After planting light irrigation is applied to settle the soil around roots.

Training and Pruning

Phalsa is usually trained as a bush. It can also be trained on head system with a single stem. The height of the single stem should be kept at 90cm to 1.0m. The plant produce shoots above this height just as in bush system. In North India, Phalsa is pruned every year during January. When it is trained as a bush the shoots are cut from the ground level by leaving 2-3cms stubs. From here many sprouts emerge on the stubs Flowering occurs on these shoots in March-April.

Irrigation

Phalsa can tolerate drought conditions fairly well after fruit harvest. For getting high fruit yield it requires irrigation at regular intervals of 20-days during April to June. No irrigation may be applied during rainy season and in dormancy.

Manuring and Fertilization

Apply 5kg of farm yard manure to each bush after pruning in January. The bushes may be applied 50 to 100 grams of urea in two parts *i.e.,* during March and April depending upon the age. High N dose lead to profuse shoots growth which is not desirable for good fruiting. When the bushes be come four years of age increase the dose to 200gm in split dose. Apply 100gm in March and rest 100g, in April at months interval.

Weed Control

Phalsa requires two hoeings. One after pruning in January

and second in April-may. If the intensity of weeds increases in rainy season, then spray Gramoxone (paraquat) @ 6-7ml/L. of water in vacant places in plantations which are trained on head system. However, no herbicide needs to be sprayed in bush trained plants. Shade of the foliage keeps a check on the weeds growth.

Fruit Maturing and Harvesting

For uniform ripening of phalsa fruits, apices of shoots may be pinched in mid-May to check further shoot growth. Fruits start ripening in the first week of June and continue for a month. Fruits should be harvested twice a week. Fruits are packed in small baskets of size 2kg or in packs. Phalsa fruit is perishable in nature, hence should be transported to the market soon after harvesting.

PLANT PROTECTION

A. Insect-Pests

1 Phalsa Hairy Caterpillar (*Eupcrotis fraternae*)

It is a serious pest of phalsa. Caterpillars feed on leaf lamina and skeletonize it completely. In severe cases whole bush may be defoliated.

Management

Spray Thiodan 35EC (Endosulphan) or Dursban 20%EC (chloropyriphos) @ 2ml/L of water immediately as the caterpillars are seen.

2. Psylla

It suck the cell sap from young leaves. The growth is reduced and fruiting is affected.

Management

Spray Rogor 30EC (Dimethoate) @ 2ml/L of water in first week of March.

3. Fruit borer

Observe the fruits in first of week of May if attack is seen only then spray.

Management

Spray Sevin 50WP (Carbaryl) @2g/L of water as soon as holes in developing fruits are seen.

B. Diseases

1. Brown spot (*Cercospora grewiae*)

Brown spots appear on the leaf lamina from June to September. The whole lamina turn brown and dry up. As a result premature leaf fall occurs. In severe case shoots may dry up.

Control

Spray Bordeaux mixture 2: 2 : 250 twice once, in June and then July OR spray Dithane Z-78 @3gm per litre of water as soon as spots are seen on leaves.

2. Powdery Mildew (*Oidium Sp.*)

Disease appears after rainy season and cause defoliation of shoots.

Control

Spray wettable Sulphur @ 2g/L of water as soon as white powdery mass appears on shoots or under the leaves.

SAPOTA

Botanical Name

Achras zapota or *Achras sapota* L

Family

Sapotaceae

Sapota commonly called chiku is an important fruits of India. Due to its hardy nature. It is the most common plant of the kitchen gardens. It is very much liked by both old and young due to its sweetness and fleshy pulp.

Origin

It originated in tropical America, from where it spread to Malaya, Philippines, India, Indonesia and Sri Lanka.

Area and Production

Sapota is cultivated over an area of 156.1 thousand hectares in India with annual production of 13.08 lacs MT. It is grown from Jammu to Dehradun along Shivalik hill. Sapota is the most preferred fruit plant of kitchen gardens in Chandigarh and Panchkula. No commercial plantation has come up in Punjab plains due to low yield per trees. Sapota is widely cultivated in Gujrat, West Bengal, Karnataka, Andhra Pradesh, Maharashtra and Tamil Nadu.

Uses

Sapota fruit is a rich source of Sugars (13-18%). Pulp contains sufficient quantity of minerals per 100 grams of pulp. Fully ripe fruits are eaten along with skin (epicarp). Skin can be removed for old and babies. Out of India Sapota is cultivated for the extraction of Chicle gum. Sapota bark has medicinal value.

Botany

Sapota plants are ever green in nature. The tree canopy has four kinds, *viz,,* erect growing; with drooping branches, spreading branches with sweet fruits and spreading branches with inferior fruits.

Tree can grow up to 8m high. It has strong trunk on which scaffolds develop at regular intervals. Fruit bearing is on new growth in axils of leaves. Flowers have 6 sepals 6 stamens which are petaloid. The fruit is berry. The fruit epicarp resembles that of potato. The fruit may be seedless or may have 3-5 black shinning seeds.

Climate

Sapota is a tropical fruits but can be grown under sub-tropical and temperate conditions just like mango. Its cultivation is successful on the coastal are. High temperatures of Punjab plains adversely affect the flowers and fruits both.

Soil

It can be grown on wide range of soils ranging from sandy loam to clay loam. It can tolerate salty conditions to a great extent.

Cultivars

Cultivars are recognized on the basis of shape of the fruit or color of the foliage. Cultivars with mellow and sweet flesh are preferred over hard fleshed ones.

1. Kalipati

As the name shows the plants of this cultivar have dark green leaves which look like blackish in colour. Branches are spreading in habit. The fruits are oval in shape. Seeds vary from 1-4 per fruit. Fruits appear singly. Trees are heavy bearer and fruits are harvested in winter.

2. Cricket Ball

The leaves are light green in colour. Fruits are round in shape and bigger in size. Pulp is gritty and sweet than kalipati. The skin (epicarp) is thick.

3. Chhatri

Trees have umbrella type canopy. The leaves are of light green in colour. Fruits are oval in shape. The bearing of trees is good.

4. Calcutta Round

The foliage is light green in color. Fruits are of large size, with gritty flesh. It is prone to leaf spot disease.

5. Pala

Trees bear oval fruits of medium size which are egg shaped. Heavy bearing trees. Fruits are borne in clusters the skin of fruits is thin with sweet flesh.

6. Dwarpudi

The tree bear small, sweet fruits resembling cricket ball.

7. Dhola Diwani

Trees have light green foliage. It bears oval fruits of good quality which are harvested in summers. Fruits are of small sized with thin skin. Pulp is sweet. Some hybrids have also been released for cultivation. These are CO-1 and CO-2 from Coimbatore and PKM-I a cross between kalipati x cricket ball from Periyakulum.

8. Propagation

Sapota can be propagated though grafting and air layering. Air layering is done during monsoons (July-August). Percentage of success in air layers can be improved by applying 100 to 200ppm of Indole butyric acid solution on the ringed bark before wrapping with moss grass and plastic sheet.

Grafting can be done on 'Khirnee' seedlings are considered better than Sapota seedlings. Seedlings of khirnee can be raised in the nursery or can be collected from forests. One year old seedlings are used for grafting. Inarching is done during spring (March) on the khirnee seedlings. Inarched plants are ready within six months of inarching.

Planting

Sapota is planted in the field in July-August in the prepared pits. The block planting can be done 8m apart in a square system of plantings. If it is to be done on light soils then planting distance may be reduced to 7m. Thus 144 to 169 plants are required for one hectare which should be booked before hand with a reputed nursery.

Flowering and Fruiting

Under tropical climate sapota flowers through out the year, with two main seasons *i.e* July-November and February-March. In sub-tropical conditions it also flowers twice in October-November and April-May. The flowers are protogynous in nature *i.e.,* stigma of the ovary becomes receptive earlier than the stamens produce pollen. Hence, Sapota is cross-pollinated crop. There is sufficient fruit set but retention is low. This may be due many reasons; particularly low auxin level in ovary just after fruits set or due to sudden change in temperatures in December (low temp) and in May-June (high temp). The fruit retention can be improved by spraying GA_3 @ 100ppm at full-bloom and fruit set. Or NAA @ 300ppm twice. It takes over 10 months for full maturity of the fruit. Sapota is a climacteric fruit and can be harvested when fruits are still hard.

Irrigation

Sapota is a hardy plant and can tolerate scarcity of soil moisture fairly well. For good growth and high yield, plants may be irrigated at an interval of 7-10 days in summer and 20-25 days in winters.

Manuring and Fertilization

After planting no fertilization is given for the first year. In the second year apply 20kg of farm yard manure and 50g urea + 60g super phosphate plus 20g of Muriate of potash. The following dose is being applied in A.P. and Karnataka.

Age of the (Years)	*Farm Yard Manure (kg)*	*Urea (46% N) (g)*	*Superphosphate (16% P_2O_5) (g)*	*Muriate of Potash (60% K_2O)(g)*
1 – 3	50	50	20	75
4 – 7	50	100	40	150
7 – 10	50	200	80	300
11 and above	50	400	160	450

These doses are applied in two halves first half in June-July and second half in September-October. Under North India conditions October fertilization can be injurious due to cold winters in the coming months.

Intercropping

In the early years of planting when sufficient vacant place is availablesome intercrops can be grown. During rabi season, grams/wheat or sarson can be grown similarly in Kharif grow pulses. Keep separate irrigation system the fruit plants. Vegetables can be successfully grown. Care should be taken that root system and scaffolds are noted damaged during ploughing. Papaya can be planted as a filter.

Training and Pruning

A strong trunk is necessary for a strong framework of the tree. Train the Sapota plants by removing the lower branches on the trunk up to 45cm. the lowest scaffold should be selected at least 50cm above the ground level. No pruning annual pruning is needed it bears on top of shoots in leaf axils.

Improving Fruit Set

Fruit set can be improved by spraying GA_3 @ 100ppm at fullbloom and after fruit set. NAA spray @ 25-100 ppm at flowering and repeated after 15 days resulted in better fruit set at Navsari.

Harvesting and Handling of Fruit

Sapota is a climacteric fruit hence, should be harvested at full maturity. It is little difficult to judge proper fruit maturity due to overlapping of flowering and fruiting. If immature fruit are harvested, these take long time to ripen and quality is always poor. It takes about 10 months to mature Sapota fruit. The mature fruits shed off the brown scaly scurf from skin. Mature fruits do not show green tissue or latex underneath the brown skin of the epicarp. A full gown tree can yield over one quintal of fruit up to 25-30 years of age of the tree.

The harvested fruit are graded into large. Medium and small sizes. The fruits are packed in wooden carts which are padded with paper cuttings, straw or banana leaves. After ripping fruits become very soft and can bot be kept for long. Ripe fruits can be stored at 2-3°C with 85-90% RH for a month or so.

PLANT PROTECTION MEASURE

A. Insect-Pests

1. Mealy Bug (*Phenacoccus icerjoides*)

It is sucking insect which causes damage by sucking cell sap. Developing fruits drop off.

Management

Apply slippery band as in mango. Spray the trees with Rogor 30EC (dimethoate) @ 2ml/L of water.

2. Fruit Borer (*Virachola isocrates*)

The infested fruits/buds produce latex and can be easily identified. For the management of borer spray the trees with Thiodan 35EC (Endosulphan) @ 2ml/L of water or Dursban 20%EC (Chloropyriphos) @ 3ml/ L of water as soon as the attack is noticed.

3. Scale Insects *(Pulvinaria psidii)*

Scale suck cell sap from leaves and young shoots. Scales are brown in colour and attack the foliage in large numbers.

B. Diseases

1. Leaf Spot

This is caused by fungus *Phaeophleospora indica.* A Number of pinkish to reddish spots appear on the leaves. Leaf spot disease is severe during rainy season.

Control

Spray Bordeaux mixture 2 : 2 : 250 during July-August. Two sprays are enough to check the disease.

2. Flat Limb

Commonly called 'fasciation'. It is caused by fungus *Botryodiplodia theobrome*. Branches get flattened. Flowers and leaves get clustered at the tip of affected branches.

Control

Prune the affected branches and destroy.

POMEGRANATE

Botony Name

Punica granatum L.

Family

Punicaeae

Pomegranate is one of the esteemed table fruit. Its fruits are known for their sweetness and fine blend of acidity. Its juice is cool and refreshing with medicinal value. The hardy nature of pomegranate make it the choicest kitchen garden fruit. Highly acidic fruits with 7-8% acidity are used to make 'anardana' (daru).

Origin

Pomegranate is native to Iran. It is commercially cultivated in Afghanistan, Baluchistan, Iraq, Spain, Russia, India, Pakistan, Myanmar, China and Japan.

Area and Production

In India, pomegranate production is 8.07 lacs MT from an area of 109.2 thousand hactare and Maharashtra state contributed a share of 75.09 per cent to the total India production of pomegranate. The major growing states one Maharashtra, Karnataka, Gujarat, Andhra Pradesh. Chhatisgarh, Himachal Pradesh, Nagaland, Orissa, Rajasthan and Tamil Nadu.

Uses

The sweet fruits are used as table purpose and acidic for making 'anardana' a condiment used for 'chatni' making. Juice is prepared and preserved for use. Wine can be prepared from sweet coloured juice. Sodium citrate/citric acid is prepared from very acidic fruits. Pomegranate shoot bark is used for body slimming. The root or rind (peel) of the fruit is used to control dysentery, diarrhoea and worm killing in the intestines. Every part of the pomegranate plant is used for one or the other human disease. Dyes are prepared from flower petals. The edible portion (grains) of fruit contain upto 78 percent water, 5% fibre, 1.6% protein, 16-18% carbohydrates (sugars). Each 100 gram of fruit flesh may contain calcium 10 mg, phosphorus 70 mg. and sufficient quantity of iron, riboflavin and vitamin C.

Botany

It grows as a shrub but can be trained on modified leader system as a small tree. Pomegranate is evergreen as well as deciduous. Shrub trained plants remain smaller than single stemmed trained as trees. It is hardy tree and can live over 40 years. The leaves have small petioles and are oval to lanceolate in shape. The shoots have thorns which originate deep from the wood. It bears very beautiful red coloured flowers. Flowers may be solitary, axillary or appear in clusters on short spurs. The calyx is persistent and tubular with 5-7 petals inserted in calyx. Ovary have many locules. Pomegranate fruit is special fruit botanically. The edible portion is aril juicy covering of seed. Both types of seed are present is different cultivars. The hard seeded and soft seeded cultivars.

Soil and Climate

Pomegranate thrives well in semiarid conditions. It can

adapt wide range of soil and climatic conditions. Deep loam to sandy loam soils are considered ideal. It can tolerate alkaline/ saline soils with 9.0 pH having lower EC than 0.5 mm hos/cm. It can grow in light soils but with assured irrigation. It requires hot and dry summers with cool winters. It is tolerant to frost and freeze fairly well very high temperature in summers and too low in winters encourage fruit cracking. Pomegranate have both types of cultivars, *i.e.* some are deciduous in winter and others are evergreen. The tree requires hot and dry climate for the production of high quality pomegranates. The evergreen cultivars do not shed leaves in very cold winters.

Cultivars

Mostly pomegranate cultivars originated as seedling selections. Some promising cultivars have also been developed through controlled hybridization. The recommended cultivars are area specific for example Kandhari and Nabha for north plains, Jaloreseedless and Jodhpur white in Rajasthan, Velludu in Tamil Nadu, Dholka in Gujrat, Basse-in-seedless, Bhagawa for Karnataka and Phule Arakta, Ganesh in Maharashtra.

Some of the promising cultivars are discussed below :

1. Ganesh

A seedling selection from Alandi made by Dr. G.S. Cheema at Pune. Plants bear profusly and regularly. It is evergreen bush and precocious. Its fruits are of medium size, with yellow coloured peel with pink blush. Average fruit size 350 g. Arils are white, transparent with little pink tings. Seeds are soft and can be eaten along with arils. The juice percentage on the basis of grain weight is 80% or more. The TSS of juice varies from 13% to 16% with very little acidity (0.3 to 0.5%). Tree can bear Ambe, Mrig and Hast bahars. One can opt for any bahar as per requirement of the area. Under North Indian conditions, it is beneficial to take Ambe-bahar, which ripens in August. To improve the fruit size, flower buds appearing after 15th April should be manually removed. Fruit yield is 12-15 tonnes per hectare.

2. Kandhari

Trees are deciduous, vigorous and upright growing. It is regular bearer with good yield per tree. It bears only Ambe-bahar (April-May flowering). Fruits are of big size with yellow coloured peel and red splash. Average fruit weight is 370 g Arils are light pinkish to deep pinkish, with semihard seeds, depending upon the dry weather and prevailing temperature. The grains have 77 percent juice on the basis of weight of grains. The juice have 13-15% TSS and well blended acidity (0.5 to 0.6%).

To get well-sized fruits of high quality, remove the flower bud appearing after 25th of April manually. There is a considerable variation in yield from area to area. However, on an average it can yield 10-12 tonnes per hectare.

3. Nabha

Tree growth is just like Kandhari and is also deciduous in nature. The rind of the fruit remain yellow in colour. The fruit size becomes larger than kandhari if thinning is done. The average fruit weight is 350 g. The grains are of white colour, with hard seeds. The grains have 75 percent juice with 13% TSS and 0.5% acidity. Yield 10 tonnes per hectare.

4. Assam Local

Under North Indian conditions tree grows upto 3 m high and 1.5 m spread. Tree is healthy. The fruits remain free from fungal/bacterial attack during the rainy season. It also bear only one crop. Average fruit weight is 220 g. The arils are small, hence juice percentage on the basis of weight of grains in 47 with 15% TSS and 0.6% acidity. Average yield per hectare only 7 tonnes.

5. Jyothi

Trees are dwarf and evergreen. Under North Indian conditions also it bear three crops, *i.e.* Ambe, Mrig and Hast bahars. Due to very low temperature in January the fruits of mrig bahar crack. Therefore it is beneficial to get only ambe bahar. Fruits remains small in size. Average fruit weight is 200 g. The rind develops a typical red colour with pinkish-reddish coloured arils. The juice percentage is 75 on the basis of grains weight. The juice have 17% TSS with very low acidity 0.3%. Average yield 8 tonnes per hectare (ambe-bahar only).

6. P.A.U Selection

It originated as a chance seedling from Ganesh and was evaluated along with other thirty cultivars at Ludhiana. The plants are as vigorous as Ganesh and evergreen. The flesh characters resemble Ganesh, but seeds are little hard. It bears big-sized fruits like kandhari. Average fruit weight 450 g. Juice percentage on the basis of grain weight is 71% with 13% TSS and 0.7 acidity. Average yield 8 tonnes per hectare.

7. Mridula

It is a soft seeded evergreen cultivar and it is cross between Gul Shah Red × Ganesh. It also bear three crops, but it is better to get Ambe-bahar only. The plants remain dwarf than Ganesh. The fruits have red coloured rind. The grains are also blood red in colour. If thinning is done in Ambe-bahar the average fruit weight is 240 grams. The juice is 78 percent on the basis of grain weight the TSS of juice 17-18% with only 0.3% acidity.

8. Basse-in-Seedless

As the name shows the grains are seedless. The plants are as vigorous as Ganesh and are evergreen in nature. The fruits are prone to fungal/bacterial attack during rains. The average fruit weight in ambe bahar is 340g. The grains are light pinkish as of Ganesh. The juice is 80 percent with 14% TSS and 0.4% acidity.

9. Ruby

The plants are slow growing, evergreen and bear all the three crops. It is a hybrid from a 3 way cross between Ganesh × Kabul × Yercand and Gulsha Rose Pink. It bears red coloured fruits of small size with red coloured arils. Average fruit weight is 200 g. Juice percentage 80 with 15% TSS and 0.3% acidity.

Some other important cultivars are, G-137 a clonal selection from Ganesh; P-23 and P-26 seedling selections from Muscat; IIHR selection, it is a selection from open pollinated seedlings; Yercand-1 and Co-1.

Propagation

Pomegranate should not be propagated through seeds, due to cross pollination, considerable variation occurs in the seedlings. Some workers say that pre-conditioning of shoots during June-July by girdling or etiolation increases the level of root promoting cofactors, which help in rooting of cuttings. The maturity of shoots used for cuttings plays a great role in the rooting of cuttings. For getting healthy and precocious plants, pomegranate should be propagated through hard wood cuttings. One season old cuttings should be prepared during first week of December. The length of cuttings should be 20-25 cms. Treat the lower ends of these cuttings with 100 ppm of IBA (Indole butyric acid) solution for 24 hours, before planting in the nursery beds for rooting. After 10 to 15 days of planting, the beds may be applied 5 litres of chloropyriphos solution @ 10 ml/L. to one square meter of the bed. Repeat this treatment 20 days after to control white ants attack. Chloropyriphos solution should be applied with a shower over the cuttings slowly so that it reaches the rootzone apply light irrigation after the treatment.

The cuttings should be irrigated lightly at an interval of 15-10 days from December to March and 7 days after wards. Aphid attacks the newly emerging foliage, which should be controlled by spraying Rogor @ 2 ml/L. of water or Nuvacuron (monocrotophos) @ 1 ml/L. More than 80 percent cuttings root and produce healthy plants. Deciduous cultivars root better than the evergreens. To improve rooting of cuttings in evergreen cultivars take only one crop, *i.e.* ambebahar remove flowers from June onward.

Planting

Pomegranate should be planted on square system of

planting at a distance of 5m × 5m apart. Ever green cultivars may be planted at 4m × 4m distance. Prepare 1 meter diameter and 1 m deep pits one month prior to planting of pomegranate. Thus one need 400 to 625 plants for planting an hectare. The best time for planting of pomegranate is December-January. The evergreen cultivars should be planted in December. Young plants are attacked by white ants in the pits also. So apply one litre solution of chloropyriphos @ 10 ml/L. of water after two months of planting to each plant.

Training

Normally no importance is given to the training of pomegranate plants. Pomegranate is left as such to grow as a bush which is not desirable. Plants can be trained as single stemed tree (*e.g.* Kandhari etc.) and 2 to 4 stemed bushes (evergreen). Multi stemed trained plants create problems later on as the stems inter mingle with each other.

Pomegranates should be trained as single stemed to get strong scaffold system. No branch should be allowed to develop from ground level to 30 cm of trunk height. Head back the leader at 1 m height to force the scaffolds to develop. Select only 4-5 well placed scaffolds on all sides of the main trunk. If the plants are trained on two or three stems, then the plant it self develop a balanced scaffold system. Remove only intermingling branches from the stems. It has been recorded that deciduous cultivars perform better as single stemed trees and evergreens with 2 to 3 stems.

Pruning

Pomegranate do not require annual pruning. The fruit is borne on short spurs as well as in leaf axils and shoot apex. Remove criss-crossing and dried branches. Some branches may be headed back during December by removing 30 percent of the growth to encourage fresh growth.

Crop Regulation

Pomegranate is a precocious bearer. Plants start bearing in

the 3rd year of planting. It continues to bear for 30 years. The evergreen cultivars flower through out the year and bear three crops. To obtain good quality of fruits and high yield. Select only one bahar as suited to the area. Remove the flowers manually for rest the bahars. Deciduous cultivars only bear Ambe bahar (flowering April-May). The evergreens flower in Ambe bahar as well as Mrig bahar (July) and Hast bahar (October-November).

The inflorescence is cyme, due to abscission of side buds and persistence of central flower, flowering appears to be solitary. Keep only one fruit at a place to get well-sized fruits. The flowers should be manually removed after mid-April when sufficient number of fruits have already set. On a mature tree select 80-100 fruits only.

Pollination

Pomegranate bears three types of flowers *i.e.* pure male with rudimentary ovary; hermaphrodite with medium style and hermaphrodite with well developed style. The percentage of these flowers varies from cultivar to cultivar and bahar to bahar. There is no shortage of pollen and pollinating agents. Sufficient number of fruits set from hermaphrodite (pin) flowers. Pollen is available at noon and stigma remain receptive for 2-3 days. Both cross and self pollination take place.

Irrigation

Apply irrigation just after transplanting the plants in December. Light and frequent irrigations should be given at an interval of 20 to 10 days from January to May and at weekly interval from May to end July. No irrigation may be given if rains comes in. Increase the interval of irrigation. After the harvest of fruit in August September. The interval of irrigation may be a month. Deciduous cultivars may not be applied any irrigation during December-January when leaf fall starts.

Intercropping

Pomegranate have short juvenile period. Plants are also

planted closely and remain bushy and spreading in nature, hence, no intercrop should be grown. When planting distance is more than 4m × 4m then intercrops can be grown for the first two to three years. Growing of vegetables and pulses should be preferred over rabi crops and fodder crops. If wheat is to be grown prepare separate irrigation system for irrigation to pomegranate plants during March-April.

Manuring and Fertilization

Manuring should be done as per nature of plant, *i.e.* evergreen or deciduous. If crop is not regulated then deciduous trees bear two crops and evergreen three crops in a year. Accordingly, fertilizer requirements increase as the number of crops in a year. As it has been recommended to take only Amb-e-bahar in North India, to get good quality of fruit. Following fertilizer doses may be applied.

Add farm yard manure, super phosphate and muriate of potash in December-January. Mix well in plant basins. Urea should be split into two parts. Apply half in February at sprouting and half in May after fruit set.

Age of the Tree (Years)	*F.Y.M. kg/tree*	*Urea (N 46%)*	*Superphosphate (P_2O_5 16%) g/plant*	*Muriate of Potash (K_2O 60%)*
1 – 2	10 – 20	50 – 100	100 – 200	30 – 60
3 – 5	30 – 50	150 – 250	300 – 500	90 – 150
6 and above	50	300	600	200

However, different fertilizer doses are given in different areas. In general a higher dose has been recommended in literature *i.e.* 1.5 kg urea, 1 kg superphosphate and 400 grams of muriate of potash per mature plant. This dose may hold good where more than one crop per year is taken.

Weed Control

Pomegranate is closely planted, hence use of a tractor to intercultivate is not desirable. Weeds should be manually removed by hoeing the basins once in December at the time of fertilization and then again in May. The left out places may have some weeds/grass growing. If need be spray Grammoxone (paraquat) @ 6 ml/L of water during July. Keep the nozzle of spray close to the weed growth to avoid drift of the chemical.

Harvesting and Marketing

Pomegranate is non climacteric fruit hence should be harvested when fully ripe. Pomegranate fruits become ready for harvesting after 4-5 months of fruit set. As the flowering continues for over two months, in the same way fruit ripening and harvesting also continue for a month or so. The fruits are harvested at full maturity, when rind has developed a typical colour of the respective cultivar. At this stage calyx at the tips dries up. Fruits should be harvested with the help of secatures.

Grading

Fruits are graded on the basis of fruit weight and size. For big markets following grades are followed :

1. Super sized : weight more than 750g, without spots.
2. King sized : 500g – 750g without spots.
3. Queen sized : 400g – 500g without spots.
4. Prince sized : 300g – 400g well coloured.
5. 12-A : 250 – 300g without spots.
6. 12-B : 250 – 300g with some spots.

Marketing

Fruits are packed as per their grade. CFB boxes are used for packaging and fruits are cushioned with paper strips/cut pieces. The fruit is transported to the markets. For small markets fruits are packed in wooden boxes of 5 to 10 kg size.

The suitable temperature for cold storage of pomegranate fruits ranges between 5-10°C, with 90-95% relative humidity. Fruits stored at lower temperatures 0-3°C develop chilling injury.

Physiological Disorder

1. Fruit Cracking

Fruit cracking usually occurs in fruits of mrig and hast bahars. It is more serious in arid zone. Cracked fruits get affected by fungii and fruits become unfit for consumption. Mrig bahar fruits crack due to wide variation in humidity and soil moisture. Prolonged drought in ambe-habar causes peel hardening and when rain comes or irrigation is applied the arils sweel and peel cracks. Some cultivars are more prone to fruit cracking. Hast bahar fruits crack due to low temperature in December and January.

Cracking can be controlled by supplying irrigation at regular intervals, this will reduce the fluctuations in soil moisture. Planting of wind break around the orchard also help in the reduction of fruit cracking. Spray of borax @ 0.1 to 0.2 percent, spray of GA_3 30-40 days before fruit ripening @ 15-20 ppm checks cracking of fruits. Cultivars Ganesh, Kandhari and PAU selection show little cracking in ambe bahar fruits.

2. Internal Break Down of Arils

Blackening or discolouration of arils in ripened fruits is a serious malady through out the pomegranate growing areas. All cultivars are affected by this disorder. No cause or remedial measure can be suggested at this stage. The fruit should be harvested at ripening stage without keeping it on the tree for more time than required.

PLANT PROTECTION

A. Insect-Pests

1. Aphid, Thrips and White Flies

These are cell sap sucking insects and attack the plants in large numbers. These insects attack the new foliage, flowers and tender fruits. A black sooty mould develop on their excreta and hinder the photosynthetic activity of leaves. The flowers and fruits drop due to excessive cell sap extraction.

Management

Spray Nuvacuron 36 SL (monocrotophos) @ 1.5 ml/L of water or Rogor 30 EC (dimethoate) @ 2ml/L of water as the buds sprout to produce young shoots. Repeat the spray if attack is seen.

2. Fruit Borer or Pomegrante Butterfly (*Virachola isocrates*)

This is a serious pest of pomegranate, throughout the growing area. Attack may be seen from flowering stage. In North India attack is from May onward caterpillars bore into developing fruits and feed on pulp just below the rind. The excreta of caterpillars and fluid oozes out of the holes in the rind. The portion of rind rot and fruits start dropping.

Management

(*i*) Collect and dump in the infected fruits.

(*ii*) Spray during May-Nuvacuron 36 SL (monocrotophos) @ 1ml/litre, repeat the spary after 15 days. After 15 days give another spray with Thiodan 35 EC (endosulphan) @ 2 ml/litre.

(*iii*) If possible, fruits may be covered with muslin cloth bags or polythene bags with small holes in it during first week of May till harvest.

3. Bark Eating Caterpillar (*Indarbela tetraonis*)

The pest bore into crotches of scaffolds. The frass of excreta and wood comes out of holes. The trees become weak and produce low yield.

Management

(*i*) Open the trees by removing criss-crossing branches.

(*ii*) Inject kerosene oil or chloropyriphos solution 50 : 50 water in the holes and plug it with mud.

B. Disease

1. Leaf Spot

Xanthomonas punicae and various fungii such as *Colletotrichum gleosporiodes*, *Fusarium fusarioides*, and *Phomopsis ancubicola* attack the foliage and cause black or brown leaf spots on the leaf lamina.

Control

(*i*) Spray Bordeaux mixture 2 : 2 : 250 twice, just after fruit set and one month later.

(*ii*) Spray tillex @ 2g/litre during July.

2. Fruit Rot

It is caused by fungus *Glomerella cingulata*. The fruits show 'off' colour at the calyx end. The affected portion of rind turns brown. Later on the area of rot increases and whole of the fruit get rotten. Fruits start dropping.

Control

(*i*) Spray Bordeaux mixture 2 : 2 : 250 as soon as spots are noticed on the fruit.

(*ii*) After 15 to 20 days of first spray, spray Banistin @ 1g/ litre of water.

3. Black Rot

This fruit rotting is caused by bacterium *Xanthomonas punicae*. With the onset of rains black spots develop on the rind of mature fruits. The intensity increases at a fast speed due to increased humidity and fruits start rotting.

Control

Spray Bordeaux mixture 2 : 2 : 250 during May, after fruit set and repeat the spray in June and July at monthly interval.

LOQUAT

Bot. Name

Eriobotrya japonica Lindl.

Family

Rosaceae.

Sub Family

Pomaceae

Loquat is also known as Japanese plum. It is being cultivated successfully under sub-tropical conditions. In North India loquat fruit ripens during end March to May, at a time when there are only few fresh fruits available in the market.

Origin

Loquat is native to China. Many good loquat cultivars also originated in China and Japan. It has become a commercial fruit of Australia, U.S.A. Hawai, France and India. U.P. occupies the first position as for as area under loquat is concerned. Loquat is highly preferred fruit for kitchen gardens in Punjab. It is also being grown on a small scale in the orchards at Amritsar, Gurdaspur Hoshiarpur and Patiala districts.

Uses

Loquat is a delicious summer fruit. It contains 60 to 70 percent pulp depending upon the cultivar and 15-20 percent seed. The pulp is composed of 88% water, 0.6% proteins, 9.6% carbohydrates, 0.5% minerals, 1.0% fibre and 55% IU vitamin A. Some preparations like Jam, Jelly and squash are made from the pulp. It is consumed as a fresh fruit. The colour of epicarp usually ranges between yellow-red. The fruit tastes sweet and sour with good flavour.

Botany

Loquat belongs to family Rosaceae and sub-family

pomaceae. The fruit is pome with bold seeds. The trees are evergreen, spreading and errect growing. A full grown tree may attain a height of 7-8 meters. The leaves are large and densely located at the ends of shoots. Shape of leaf is lanceolate with dentate margins. The upper surface is green but lower surface rusty tomentose. The flowers appear on inflorescences arising terminally on new shoots.

Flowers have 5 sepals, 5 petals, 20 stamens and 5 pistils joined at the base. Shape of the fruit varies from round to pyriform. Flesh firm and melting. The colour of pulp varies from whitish to deep orange. The seed size and number varies from fruit to fruit and cultivar to cultivar. Ovary is 5 celled with 20 ovules. Some ovules abort and only 2 to 5 ovules turn to develop into seeds in a fruit. Seeds are bold and have brown tasta. The edible portion is thalamus just like other pome fruits.

Climate and Soil

Loquat has adapted well to the harsh climate of North India. It is performing well in tropical climate and lower hills, where temperature rarely falls below 0°C. It cannot tolerate severe frost in winter. It is tolerant to drought and heat. It likes warm and dry climate at fruit ripening time.

Loquat grows well on fertile clay loam soils. It can be grown in loam as well as on sandy loam soils. Waterlogged conditions are harmful to the plants.

Cultivars

Important loquat cultivars have been collected and evaluated at Punjab Agricultural University, Ludhiana.

1. Golden Yellow : It produces medium-sized egg shaped fruits, with attractive golden yellow colour. The fruit flesh is also yellow coloured, which tastes sour-sweet. Each fruit contains 4-5 medium sized seeds. It ripens in end March.

2. Improved Golden Yellow : The fruit colour is similar to golden yellow cultivar, but shape is pyriform. The pulp is orange coloured, crisp and sub-acid. Each fruit may contain 3-5 seeds. It also ripens in end of March.

3. Pale Yellow : The fruits are large slightly conical to roundish in shape. Colour of fruit is pale yellow with white flesh. The pulp tastes sour-sweet. There are 3-4 seeds per fruit. It ripens in end April.

4. Tanaka : The fruits are orange coloured, ovate and small in size. Fruit skin has pubescence. The pulp tastes sour-sweet. The fruit have 2-4 seeds of medium size. The fruit ripens in end April.

5. California Advance : This cultivar originated in California. It produces medium-sized fruits of conical to roundish shaped. The peel of fruit is of yellow coloured with creamy white flesh. Each fruit may have 2-3 seeds. It also ripens in end April.

6. Improved Pale Yellow : This cultivar originated from pale yellow and have well sized fruits of pyriform shape. The skin colour is pale yellow. The pulp is thick and creamy. Taste sour-sweet. Fruit has 3-4 seeds. Ripens in end April.

Propagation

Seeds are sown in the nursery beds after their immediate extraction from fruits. Seeds germinate soon and become seedlings graftable within a year. Don't let the seeds dry after extraction. The seedlings are in arched during July-August. Air layering can be done but success is limited. Budding can also be done in February. There is need to standardize the method and time of propagation, to achieve high percentage of success.

Planting

It can be planted in February-March and August-September. The planting distance varies from 6 meters to 7 meters, apart depending upon the soil fertility level. Thus one hectare shall accommodate 256 to 196 plants.

One metre deep and one metre diameter pits should be prepared at least one month prior to actual plantation. Lindane 30g or two litres of chloropyriphos solution @ 10/L may be applied to each pit to check white ants.

Irrigation

Loquat is hardy plant and can tolerate drought for some time. Irrigation should be applied at longer intervals of 20-25 days during non-fruiting time and interval maybe reduced to 20-15 days during fruit development. No irrigation should be applied during rainy season and winters.

Interculture

Keep the plant basins clear of weeds through out the year by hoeing. Rabi crop should not be grown in loquat orchards since, fruit is harvested in March-April. In Kharif pulses like Moong, Mash, peas and crop like Toria can be grown.

As usual full farm yard manure, super phosphate and muriate of potash should be applied in September. Half urea should be applied in October and rest half in January. The manures and fertilizers should be mixed well uniformly in the tree basins.

Manuring and Fertilization

Age of Tree (Years)	*Farm Yard Manure kg/tree*	*Urea (N 46%) g/tree*	*Superphosphate (P_2O_5 15%) g/tree*	*Muriate of Potash (K_2O 60%) g/tree*
1 – 3	10 – 20	100 – 200	100 – 200	100 – 200
4 – 6	30 – 40	300 – 400	300 – 400	250 – 300
7 – 9	40 – 50	500 – 750	350 – 500	350 – 400
Over 10 year	50	800	600	500

Flowering Fruiting

Loquat flowers from September to January under sub-tropical conditions. The panicles producing flowers early in the season also produces well sized fruits, in comparison to those appearing late.

Loquat cultivars are self-incompatible or partially self-fruitful. Therefore, do not plant only one cultivar in an orchard California Advance is a good pollinizer for Golden yellow and pale yellow cultivars. Two to three cultivar planted together give high yield.

Training and Pruning

Loquat is also trained on modified leader system of training. However, plants are also left to grow of their own and automatically develop a sound scaffold system. The tree trunk should be kept without any scaffold upto 50 cm of trunk height. The fruits are harvested as whole panicles and small portion of shoot. This much pruning is sufficient to get new growth. If some branches are to be removed, these should be removed in June.

Fruit Maturity and Harvesting

Plants start bearing after three years of age. Fully matured and well coloured fruits with TSS above 9% should be harvested. All the fruits in a cluster almost ripen uniformly. While harvesting a small portion of the shoot should also be harvested with a sharp secature. Separate the clusters as A and B grade. The culled fruits of poor quality may be sent to be local/small market.

Packing

The graded fruits are properly packed in boxes. The boxes may be wooden or card board. Well sized unbruised fruits should be packed in A grade and small sized good fruits in B grade. Rest of the fruits may be packed separately. The paper cuttings can be placed at the bottom of each box and for providing cushion to different layers. Wooden boxes of 14 kg size should be used for sending the fruit to near by markets.

PLANT PROTECTION

A. Insect Pest

1. Bark Eating Caterpillar (*Inderbela quadrinota*)

Bark eating caterpillars cause huge damage to loquat trees.

Caterpillars feed on the bark and make tunnels in the trunk. Due to tunneling, girdling is caused, which may kill the plant.

Management

Clear the holes/tunnels with wire, inject kerosene oil or chloropyriphos solution 50 : 50 water.

2. Fruit Fly (*Dacus dorsalis*)

At fruit maturity fruits are attacked by fruit fly. On hatching maggots bore into the fruits and feed on the pulp. The infested fruits become unfit for human consumption.

Management

It is difficult to control the fruit flies, once they enter the fruits. One can reduce the attack.

(*i*) Hoeing of the tree basins should be done to expose the pupae to their natural enemies.

(*ii*) Collect the infected fruits and dump in deep pits.

(*iii*) To repel the fruit fly spray Thiodan 35E (endosulphan) @ 2 ml/L. of water in first week of February and repeat the spray in the last week of February or spray Sevin 50 WP (carbaryl) @ 2 g/L. of water twice.

B Diseases

1. Shoot/Fruit Blight and Bark Canker (*Phoma glumerata sp.*)

The canker appear on bud scars, twigs or in crotches. Small circular brown spots appear around a leaf scar or superficial wound. As the canker enlarges the centres become sunken with the surrounding healthy bark. The fungus perpetuates itself on the trees in bark cankers.

Control

Remove the cankers and decorticate along with 2 cm of

healthy bark. Apply Bordeaux paste on the cut ends and wounds. Spray Bordeaux mixture 2 : 2 : 250 twice at an interval of one month.

2. Collar Rot

It is caused by *Phytophthora* species. Some are of the view that it is caused by Diplodia natalensis. The fungus produces canker from ground level to point from where scaffolds emerge. The rot girdless the trunk during 2-3 years. Affected trees flower profusely. The foliage become yellowish green. The tree show wilting and ultimately dry up completely.

Control

(*i*) Avoid flooding the orchard.

(*ii*) Uproot the infected trees and destroy.

(*iii*) Scrap the infected portion and cut some healthy bark also. Apply Bordeaux paint. Spray the trees with Bordeaux mixture 2 : 2 : 250.

3. Root Rot/White Rot

It is caused by a fungus *Polyporus palustrisis*. The affected trees show symptoms of wilt during early leaf fall and increased fruit set. The fruiting bodies appear, when the rot is fairly well advanced.

Control

(*i*) Locate the infected trees at the early stage, by examining the roots and root collar region of trees. Dig out decayed roots and cut them completely right from the collar region.

(*ii*) Treat the cut ends with disinfectant solution. Then apply Bordeaux paste. Drech the soil from where roots have been dug out, with Bordeaux mixture 2 : 2 : 250.

AONLA

Bot. Name

Emblica officinalis Gaertn.

Previous Name

Phyllanthus emblica L.

Family

Euphorbiaceae

Aonla or amla or Indian gooseberry is an Indian fruit. It is a hardy fruit crop of commercial importance. It can tolerate alkaline soil conditions, with EC 8 ds/m. Tree is a prolific bearer, even with little care. High amount of Exchangeable Sodium Percentage (ESP) can disturb plant metabolism, which has adverse effect on the growth of aonla plants.

Origin

Aonla originated in central India. From India it spread to neighbouring countries, China, Cylon and Malaysia.

Area and Production

It grows well from tropical parts to higher hills and areas with severe winter, (free from frost) where it grows as a deciduous tree. In severe frost prone areas the trees get killed. It is commercially cultivated in Partapgarh, Varanasi Sultanpur, Agra and Kanpur area of Uttar Pradesh. Its cultivation is being taken up in a big way in Maharashtra, Gujrat, Punjab, Haryana, Rajasthan and in Southern states. In India it occupies an area of 50 thousand hectares with an annual production of 1.75 million tonnes. In Punjab it is being grown on 217 hectares only.

Uses

Aonla is the richest source of vitamin C among fruits. It contains more than 600 mg of ascorbic acid per 100 grams of pulp. It also contain carbohydrates 14%, fibre 3.4%, iron 1.2%, vit B_1 30 mg per 100 g of fruit. Most of the fruit is used to prepare 'Muraba', chutney and pickle. It is used in the preparation of Chavanparash and other Ayurvedic medicines.

Botany

Under sub-tropical conditions, tree is deciduous in nature. A herb Bhoomi aonla (*Phyllanthus niguri*) also belong to family Euphorbiaceae. The herb is used to cure hepatitis (Jaudice). The bark of the trunk is grey in colour. The tree is spreading in nature. Full grown tree can have a height of 8-9 meters depending upon the soil fertility. It is monoecious plant having male and female flowers separately on the same panicle. Proportion of male flowers is very high. Male flowers appear in clusters in leaf axils and female on the upper end of few branchlets.

Climate and Soil

Aonla is primarily a sub-tropical plant but its cultivation is being successfully carried out in tropical climate too. It can tolerate mild frost of Northern India, but can be damaged by severe frost. Grafted plants are more prone to the frost injury in comparison to seedlings. Trees can tolerate hot and dry winds and drought conditions for long. Sufficient humidity is required for the growth of fruit during July-August.

Aonla can be grown on a variety of soils, ranging from foot hills to sandy loams and clay loams. It can be planted on alkaline soils (6.5-9.5pH) too. On heavy soils trees make good growth and bear profusely.

Cultivars

Earlier few cultivars were being cultivated. Predominantly small fruited seedling, Banarasi, Francis (Hathijool) and Chakaiya were grown. Due to draw backs in these cultivars, new hybrids have been released.

1. Banarasi : Fruit are large in size, light yellow at full maturity. Fruits are preferred for Muraba making average fruit weight 30g with light stone. The TSS of juice is 10%. It contains ascorbic acid 660 mg per 100 g. of pulp. Fruits have poor shelf life.

2. Chakaiya : Trees are somewhat regular than Banarsi. It also bear heavy crop. Fruits are smaller than Banarsi but are fibrous. Average fruit weight 28 g with 3g stone. TSS 10.6 which increases in post harvest stage. It contains 793 mg of ascorbic acid per 100 g of fruit.

3. Francis (*Hathijool*) : It also originated in Uttar Pradesh. The fruit size is largest of all cultivars. The bearing is less. The fruits suffer from necrosis. It is preferred for Muraba preparation.

4. Kanchan (NA-4) : It originated as a chance seedling from Chakaiya. It is heavy and regular bearer, with medium sized fruits (30g per fruit). It contain very high ascorbic acid (785 mg/100g of fruit). It is preferred by industries for pulp extraction.

5. Krishna (NA-5) : It originated as chance seedling of Banarasi. The fruit colour yellowish with red-blush. Its pulp is fibreless. Fruits are just the size of Banarasi but with heavy and bold stone. TSS of pulp is 11.0%. It contain 824 mg ascorbic acid per 100 g of fruit.

6. NA-6 : A seedling selection from Chakaiya. It is a prolific bearer. The fruits have little fibre.

7. NA-7 : It is a seedling selection from Francis. The tree is precocious bearer. Plants come into bearing after 3 years of age. It also contains very high amount of ascorbic acid (977 mg/100 g of fruit). Average fruit weight is more than 40 g. TSS of pulp 12%. Other promising new varieties are NA-10, Anand-1, Anand-2, Lakshmi-52, BSR-I.

Propagation

Aonla can be propagated through seed. For vegetative propagation seeds are extracted from desi aonla fruits during December. To improve the seed germination the seeds are soaked in GA_3 solution @ 500 ppm of 24 hours. Seedlings are patch budded when these become pencil thick, during May-June. The seedlings can also be raised in Polythene bags (15 × 30 cm) filled with a mixture of soil + F.Y.M. The inferior seedling trees can also be topworked with a promising cultivar.

Head back the desired scaffolds to keep 30 cm stubs, during February or March treate the stubs with Bordeaux paste. The new shoots emerging on these stubs are patch budded in May-June. The sprouted buds only are allowed to develop as shoots to form new scaffold system. Remove unwanted sprouts on the stubs.

Planting

The budded plants should be planted in field during August or in January-February at a distance of 7-8 m apart on square system of planting. The pits should be prepared at least a month before the planting is done. Pits of 1m deep and 1m diameter are dug and filled with top soil + farm yard manure. Aonla planting should be done with more than two cultivars in alternate lines. This will improve the yield per plant. To plant one hectare one may require 196-144 plants.

Training and Pruning

Aonla has drooping habit of branches. The bearing is heavy on tender and lengthy shoots. Due to fruit weight branches tend to break. To avoid limb breakage the trees should be trained on modified leader system of training. Lower branch should be allowed to develop at 75 cm to 1m high from the ground level only allow 4-5 scaffold branches and remove the leader after third or fourth year of planting.

Aonla trees need no annual pruning for fruiting. Remove only dried, broken branches, immediately after fruit harvest. Shoots appearing as collar suckers from the ground or rootstock trunk-below the graft union should be removed as and when these appear.

Manuring and Fertilization

Aonla plants bear heavy crop hence may be applied following doses of fertilizers and manure every year.

Age of Tree (Years)	*Farm Yard Manure kg/tree*	*Urea (46% N) g/tree*	*Superphosphate (16% P_2O_5) g/tree*	*Muriate of Potash (60% K_2O) g/tree*
1 – 3	10 – 20	100 – 300	150 – 450	40 – 120
4 – 6	25 – 40	400 – 600	600 – 900	160 – 240
7 – 10	40 – 50	700 – 100	1050 – 1500	280 – 400
10 and above	60	1000	1500	500

Apply full dose of farm yard manure, superphosphate, muriate of potash and half urea in January-February. The rest half urea dose should be applied in August. Green manuring should be carried out in alkaline soils to improve organic matter of soil.

Irrigation

Aonla is considered to be a hardy fruit plant. It can tolerate drought conditions for long time. Newly planted plants in January-February may require little irrigation at the time of planting and then at 20-10 days interval as the temperature rises. The bearing plants may be irrigated at monthly interval from September to December. Modified basin system of irrigation should be followed.

Inter Cropping

During juvenile period wheat, legumes or vegetables can be grown judiuously. Tree/plant trunks should be saved from injury. Leave a space of one meter diameter around the plants uncultivated. After the plants attain 5 years of age no crop should be grown. The basins should be kept free of weeds, manually, throughout the year.

Harvesting and Marketing

Under North Indian conditions, aonla matures in winter December-January. Fully mature fruits should be harvested by avoiding any injury to be the fruits. No shaking of trees for fruit harvest. Shaking spoils the fruits at the ground and also many branches get broken. A mature tree can yield over one quintal of fruit per tree.

At present marketing is done usually in gunny bags, which is not desirable. The harvested fruits should be placed in shade. Ladders should be used to harvest the fruits. The fruits should be packed properly in wooden boxes. Boxes should be properly nailed and transported to the desired markets.

PLANT PROTECTION

A. Insect Pests

1. Bark Eating Caterpillar (*Inderbela quadrinotata*)

The pest causes damage by eating the bark and boring holes into the stem. Due to bark eating shoots and main trunk get girdled and affect shoot/trunk dries up. Some times full grown tree dries up.

Management

(*i*) Spray Dursban 20 EC (chloropyriphos) @ 4ml per litre of water as soon as the attack is noticed.

(*ii*) Remove webbing from the holes and inject kerosene oil. Plug the holes with mud.

2. Mealy Bug (*Drosicha Sp.*)

The nymphs and female bugs climb up the shoots and suck cell sap. Mealy bug is becoming a problem for aonla cultivation.

Management

(*i*) Prevent the nymphs from crawling upto the tree trunks by applying slippery band at the base of the trunks. Lower ends of the alkathene sheet should be covered with soil and upper ends tightly tied.

(*ii*) The collected nymphs can be killed physically or splayed with Methyl Parathion or dusted with 2% follidol.

B. Disease

No disease has been reported from any Aonla growing

area. However, shoot drying has been observed in isolated plants in the foot hills of shivaliks.

Control

Spray Bordeaux mixture 2 : 2 : 250 during rainy season to fully over the foliage and interval scaffolds/branches.

13

SOME PREPARATIONS

A. Bordeaux Mixture (2 : 2 : 250)

Copper sulphate ($CuSO_4\ 5H_2O$) = 2 kg.

Unslaked lime (quick lime) = 2 kg.

Water = 250 litre

Copper sulphate and unslaked lime should be of good quality. Copper sulphate should be 98 percent pure with blue colour. Poor quality copper sulphate have whitish blue colour. This is due to oxidation of copper sulphate. As the colour changes from blue to whitish the quality is lowered. The Bordeaux mixture prepared by using oxidated copper sulphate shall not be as effective as that prepared from blue copper sulphate. The quick lime should also be good quality. It should not have much of dust (lime). The dust lime is hydrated lime dust. This is not to be used for the preparation of Bordeaux mixture. It both the ingredients are of high quality, will form a good Bordeaux mixture.

Dissolve 2 kg of copper sulphate in 50 litres of water. Pure copper sulphate easily dissolves in water at a normal temperature. During winters the copper sulphate should be placed in a muslin/gunny sac and suspended in water 125 litres. The container should be of plastic. Similarly put 2 kg unslaked lime in 50 litres of water. Mix well to form a solution add some more water to make 125 litres. Now mix the two solutions by straining into a third vessel which may contain 250 litres of solution. Mix the solution well before use. Stir the mixture all the time while pouring. Stir and strain again when pouring into sprayers.

Testing of Bordeaux Mixture before Use

(*i*) It is very important to test the mixture for its pH before actually spraying is done. A excess of copper sulphate shall give light brown colour on the litmus paper and show low pH as 5 to 6. Similarly if lime is in excess it will show blue color and range of pH may be 9 to 10 which will be alkaline. Both the pH are harmful to the foliage. Usually if the ingredients are of good quality the mixture shall be alkaline with 9-10 pH. Hence, it is advised to adjust the quantity of unslacked lime which may range from 1.5 kg to 2.0 kg. the pH on the colour scale should match with 7 to 8 jamun colour.

(*ii*) The other method to determine the excess of copper sulphate is to immerse a knife blade in the upper layer of the liquid. A deposit of copper will form on the metal. The mixture is unsafe. Add more unslaked lime to increase the pH.

(*iii*) A more accurate test can be made with ferrocyanide of potassium. Add a few drops of Bordeaux mixture to little ferrocyanide in a small saucer. If no change occurs, the mixture is safe, but if it becomes reddish brown too much copper is present. Neutralise the solution before use.

Note : It has been observed that if lime is good quality then the mixture always is of alkaline in nature.

Bordeaux mixture should not be sprayed on a rainy day or on exceptional hot day when plants show temporary sign of wilting.

B. White Wash Mixture for Tree Trunks

White washing of tree trunks is important practice in orcharding. White washing should be done twice in a year. First wash should be in April and second in September.

The following ingredients are required.

Lime dust (slaked lime) = 25 kg

Copper Sulphate	= 500 gm
Gum Suresh	= 500 gm
Water	= 100 litre

Dissolve the Suresh in hot water before putting it in to the lime mixture. In a plastic container or in a pitcher put lime and sufficient quantity of water. Then put copper sulphate in 5 litres of water in a separate container. Mix both the solutions and complete the 100 litres of mixture by adding rest of the water. Mix thoroughly and now pore dissolved Suresh, before use. Mixture should be stirred each time when taken for white washing of stems/trunks.

C. Bordeaux Paste

It is a very important paste to be used on small and major cuts when trees are pruned or broken limbs are prunned. For preparing paste dissolve 2 kg of copper sulphate in 15 litres of water. Take another 15 litres of water and put 3 kg of quick lime (un slaked lime). Mix well both the concentrated solutions by pouring one solution to the second. Stir well to form a paste. This paste is applied on the cut ends with a brush immediately after the cut is made.

D. Bordeaux Paint

Take the following ingredients.

Monohydrated copper sulphate	= 1 kg
Hydrated lime dust	= 2 kg
Boiled linseed oil	= 3kg

Prepare the monohydrate copper sulphate by placing it on a frying pan (tava) over the fire. The blue crystals of copper sulphate start lossing blue colour slowly. The crystals may be stired over the hot tawi continuously till whole the blue copper turns into white amorphous powder. Copper sulphate should be powdered with a pestal and mortal. Mix both the powder forms of lime dust and white copper sulphate (monohydrated).

Then pore the boiled Lin seed oil to the mixture. Mix the three constituents to form a homogenized paint. The paint is applied on bruised trunks and major cuts with a brush. It can be stored in a vessel for future use. Bordeaux paint makes the painted surface impervious to water/rain. Paint protects the wounds from wood-rotting fungi. Paint protects the wounds for a year.

E. Wound Disinfectant Solution

Mercuric chloride = 1 gm

Methylated spirit = 250 ml

Water = 750 ml

Prepare the solution in a glass jar/vessel. The wounded surface should be treated with this solution with the help of a cotton swap. After disinfection of the wound the Bordeaux paste or Bordeaux paint may be applied.

F. Mashobra Paste

Mix the following ingredients and dissolve the mixture in 5.50 litre of water.

Lanoline = 225 gm

Stearic acid = 12 gm

Marpholin = 150 gm

Streptocycline = 25 g.

FIGURES

Frost damaged mango plant

Kinnow on trifoliate orange rootstock

Rejuvenation of old mango tree

Top working 20 Year old Amrapali in bearing

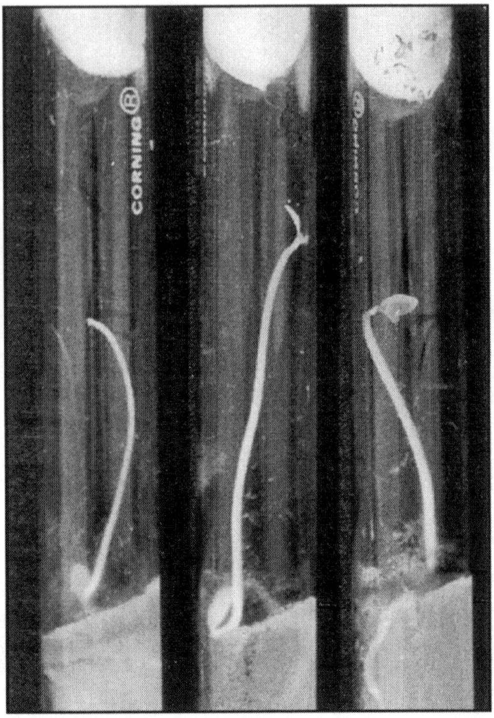

Embryo culture in M.S. media

Prunning of Ber tree

Pear rejuvenation

Pear orchard after rejuvenation

Incompatibility of pear quince rootstock

Mound Layering

Basse-in-Seedless

Plant prepared throught mound layering

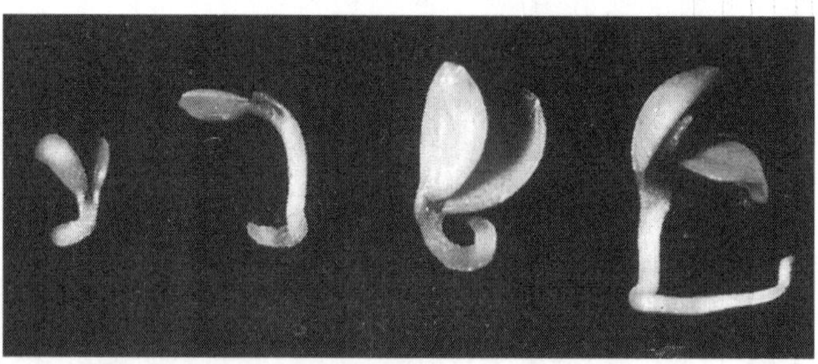

Jatti Khatti embryos from a single seed in various stages of germination. The vigorous zygotic embryo at the mycropylar end is on the extreme right.

**Fully grown guava trees segment irrigated–
No inter crops grown**

Segment irrigated fully grown pear trees

Young Kinnow plantation irrigated through separate channel system – Wheat as intercrop

Drip irrigated Kinnow and Litchi plantation infoot hills near Chandigarh

Unwanted stionic combination of Kinnow

Baramasi lemon in bearing

Pomegranate plantation

Jatti Khatti embryos in a germinating seed showing radicle protusion.

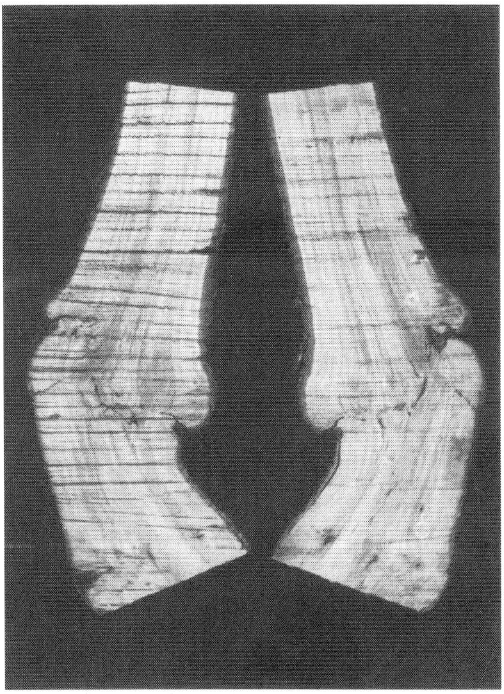

Selection showing separation of xylem and phloem tissues at union level

Inter cropping in Orchard-Garlic is being grown

INDEX

D

G